Creating Your Own IP Networks!

By

Warren Romanow

authorHOUSE™

1663 LIBERTY DRIVE, SUITE 200
BLOOMINGTON, INDIANA 47403
(800) 839-8640
WWW.AUTHORHOUSE.COM

First published by AuthorHouse 04/28/05

ISBN: 1-4208-1803-1 (sc)

Printed in the United States of America
Bloomington, Indiana

This book is printed on acid-free paper.

Table of Contents

Creating Your Own IP Networks

This book was designed to be exceptionally thorough in all aspects of voice and data communications. It takes you from the inception of the telephone to current standards based operations.

Will you need to know all of this information to create your own IP network? If you are setting up your own carrier based network, then yes. However, for your home, office, VoIP, and network based applications - no. This book contains more information than you will ever need in order to establish your network goals. However, it is accurate and it will allow you to become skilled in all aspects of the media with which you have chosen to learn.

Therefore, with this in mind, should you just wish to create a network based on my own personal recommendations, you may just want to skip ahead to chapter 9, where the entire project is laid out for you, step by step.

Some simple rules;

♦   **Before starting your project, know the basics**   ⇓

♦   **Use the appropriate proven products and services**   ⇓

♦   **Understand the process from the beginning to the end**   ⇓

♦   **Install an incredible system that saves you time and money**

Should you find any errors in this book, they are there for a purpose.

Some people enjoy looking for them, *and we aim to please!*

Please feel free to e-mail me with your questions, comments and suggestions @

ipnetworks@charter.net

Dedication

Creating Your Own *IP* Net*works* is dedicated to my children; Meghan, Breanne, Avrie, Pierce and Marlee who inspire me daily with their observations of life, as we should all view them through the eyes of a child. *They also remind me just how much damage you can do to your body and still not have to go to the hospital.*

To my wife Tracy who's never ending love and support serves as a continual beacon for direction and inspiration. I honestly do not know where I would be without her today. She is truly the love of my life! Were it not for her, this book may have never seen the light of day. *I owe her my life, and she owes me all of the hair that I have lost over the past 12 years. I am looking forward to the next 50.* Tracy, also owner of Her Vision Photography created and designed the wonderful cover for this book.

To my sister Lee, who is always there to help whenever I have needed her, but most importantly for being there when she wasn't required. She has an uncommon love and support that is genuine and kind. *I would write more thoughtful words about her, but she bent my thumbs back to my wrist when I was in the 4th grade, and I'm just now regaining mobility.*

Finally, to my mother who has demonstrated throughout my 44 years her ability to live the hardest job in the world, motherhood. I have never witnessed anyone who has done it better, with more love, empathy and dedication. I strive daily to be as good a parent to my children as she has been to me. *To date, I am still behind.*

Chapter One
Public Switch Telephone Networks

1)   Public Switch Telephone Networks

    a) Overview of the PSTN and Comparisons to Voice over IP

    b) PSTN Basics

        i)   Analog and Digital Signaling

        ii)   Digital Voice Signals

        iii) Local   Loops,   Trunks,   and   Inter-switch Communications

        iv)   PSTN Signaling

    c) PSTN Services and Applications

        i)   Standards-Based Packet Infrastructure Layer

        ii)   Open Call-Control Layer

        iii) VoIP Call-Control Protocols

        iv)   Open Service Application Layer

        v)   New PSTN Network Infrastructure Model

        vi)   Conclusion

Ever since the first voice transmission on wire was invented by Alexander Graham Bell in 1876, the Public Switched Telephone Networks (PSTN) has been in the process of continuous development. Before discussing the current scenario of the PSTN, its components and future growth, let us first understand its origin and contributing technologies. In this chapter we will start at the beginning of PSTN technology, and then gradually move on to the reason for the existence of the PSTN in the present technological perspective, basics of PSTN, its applications and various services. In the later sections of this chapter, we will discuss various ways of improving PSTN, its technologies, emerging trends and future.

Overview of the PSTN and Comparisons to Voice over IP

History of PST

In 1876, Alexander Graham Bell sent the first voice transmission through a ring-down circuit[1]. Gradually, this concept evolved from a one-way voice transmission, which allowed only one user to speak – to a bi-directional voice transmission, in which both the users could speak. This bi-directional voice transmission system required a carbon microphone, a battery, an electromagnet, and an iron diaphragm along with a physical cable between each location of the users. It should be noted, at that point in time, there was no concept of dialing a number to reach a destination.

---

[1]   A ring-down circuit means that there was no dialing of numbers; instead, a physical wire connected two devices. Basically, one person picked up the phone and another person was on the other end (no ringing was involved).

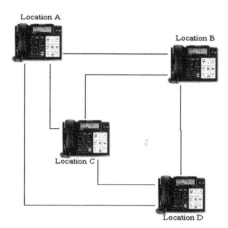

Figure 1.1

Fig: 1.1 further illustrates the beginning of the PSTN. As it is clearly shown in the figure, in this network of four-telephones a physical cable exists between each location. This concept is considered to be neither cost-effective nor feasible. In this setup, for the purpose of determining the number of lines required, you will have to denote every user as 'N' and then use the following equation: $N \times (N-1)/2$. For example, if you want to set up a network of 10 people, you will need at least 45 pairs of lines running into your house.

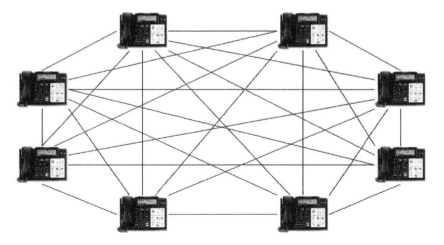

Figure 1.2: Physical Cable Between All

3

Considering this network from a broader perspective, there is a major constraint in terms of connecting a physical cable between everyone on Earth and the associated costs. Another network was developed to meet this limitation, which could map many phones with each other using a device called a switch. Switches enabled the facility to connect many phones using only one cable to the central office instead of seven.

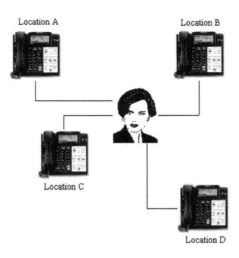

Figure 1.3: Centralized Operator: The Human Switch

In the beginning, the work of a switch was done by a telephone operator who was responsible for asking callers where they wanted to dial and then manually connecting the two voice paths. *Figure 1.3* clearly describes how the 4-phone network would look today with a centralized operator to switch the calls. With technological advancements, electronic switches have now replaced the human switch/telephone operator).

PSTN Basics

The PSTN in itself is a vast technology that incorporates numerous other related technologies. While it is impossible to cover each and every component of the PSTN in this section, this chapter covers some of the most important components that constitute the PSTN. The following sections discuss the transmission of voice across a digital network and basic circuit-switching concepts.

Analog and Digital Signaling

Each and every voice that we hear, including human speech is in analog form. In the initial phases the basic telephone network was entirely based on an analog infrastructure only. However, this analogous infrastructure too has several constraints due to its inefficiency in handling line noise[2]. In the early days, telephony networks used amplifiers to enhance the signal of analog transmission. This not only helped in amplifying the voice but also increased the noise in the line as well.

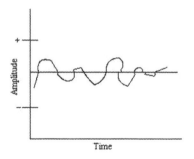

Figure 1.4: Analog Wave form

Figure 1.4 shows a graphical representation of voice with the help of an oscilloscope. Technically speaking, analog communication is a mix of time and amplitude. However it faces disturbances due to line noise, which distorts the analog waveform and causes disturbed reception. *Figure 1.5* shows that an amplifier simply amplifies the distorted signal, without cleaning the signal during the amplification

---

[2]   Line noise is normally caused by the introduction of static into a voice network.

process. This process of going through several amplifiers with one voice signal is called *accumulated noise*.

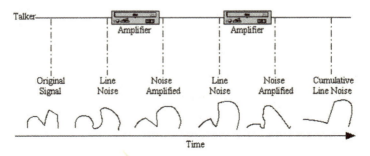

Figure 1.5: Analog Line Distortion

In different types of digital networks, line noise is not much of a serious concern as repeaters not only amplify the signal, but also clean it to its original condition. This is possible with digital communication because such communication is based on 1's and 0's. Hence, as shown in *Figure 1.6*, the repeater (a digital amplifier) decides whether to regenerate either a 1 or a 0.

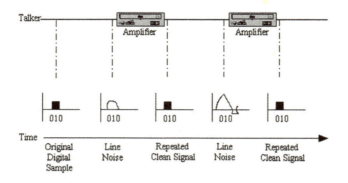

Figure 1.6: Digital Line Distortion

Ergo, a clean sound is maintained when signals are repeated. After the benefits associated with this digital representation were unveiled, the pulse code modulation (PCM) became the standard of telephony network.

Digital Voice Signals

PCM is the most common method of encoding an analog voice signal into a binary digital stream of 1's and 0's. Nyquist theorem, which basically states that you achieve good-quality voice transmission if you sample at twice the highest frequency on a voice line, forms the basis of every sampling technique.

The process of PCM is as follows:

> ➤ Analog waveforms are made to pass through a 'voice frequency filter' to filter-out anything over and above 4000 Hz. These frequencies are filtered to 4000 Hz in order to limit and check the level of cross talk in the voice network. To achieve top-quality voice transmission while using the Nyquist theorem, you are required to process 8000 samples per second.

> ➤ Further more, the filtered analog signal is sampled at the rate of 8000 times per second.

> ➤ The waveform is converted into a discrete digital form after getting sampled. This sample is represented by a code that indicates the amplitude of the waveform at the very moment the sample is taken. The telephony form of PCM uses 8-bits for the code and a logarithm compression method that assigns more bits to lower-amplitude signals.

If 8-bit words are multiplied the by 8000 times per second, the net result comes out to be 64,000 bits per second (bps). The telephone infrastructure works on 64,000 bps (or 64 Kbps). Two basic variations of 64 Kbps PCM are commonly used:

1. μ-law[3], the standard used in North America;

2. a-law, the standard used in Europe.

Both the methods are similar in a way that they use logarithmic compression to pull off 12 to 13 bits of linear PCM quality in only 8-bit words. However, they differ in relatively minor details. The μ-law method scores a slender advantage over the a-law method because of a low-level signal-to-noise ratio performance.

Local Loops, Trunks, and Inter-switch Communication

The telephone infrastructure commences with a simple pair of copper wires invading your home. This physical cabling is known as a local loop. The local loop physically connects your home telephone to the central office switch (also known as a 'Class 5 switch' or 'end office switch') by this very local loop itself. The communication path, better known as 'the phone line between your home and the central office switch' normally runs over the local loop.

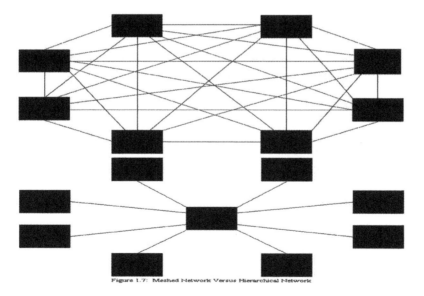

Figure 1.7: Meshed Network Versus Hierarchical Network

---

3   When making a long-distance call, any μ-law to a-law conversion is the responsibility of the μ-law country.

Trunk is the name given to the communication path between several central office switches. Just like the high cost associated with the placement of a physical wire between your home and all the other houses you desire to call, a physical wire connecting every central office switch is also not cost-effective. *Figure 1.7* clearly depicts that a meshed telephone network is not as scalable as one with a hierarchy of switches.

Switches are currently deployed in hierarchies, and end office switches (also known as central office switches) interconnect to tandem switches (or Class 4 switches) through trunks. Higher-layer tandem switches are connected with local tandem switches. *Figure 1.8* illustrates a typical model of switching hierarchy. Central office switches repeatedly get connected with each other directly.

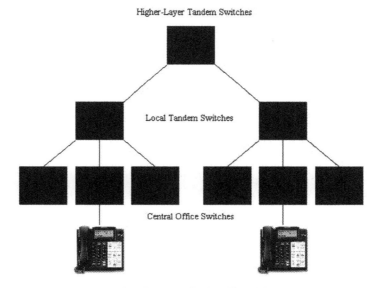

Figure 1.8: Circuit Switching Hierarchy

Where direct connections occur between central office switches, greatly depends on the calling patterns to a large extent. A dedicated circuit is positioned between the two switches to offload calls from the local tandem switches if enough traffic is occurring between the two central office switches. Some portions of the PST Network use as many as five levels of switching hierarchy in order to process calls effectively and efficiently.

PSTN Signaling

The following two types of signaling methods usually generate various transmission media:

1. **User-to-Network Signaling:** An end user communicates with the PSTN on a user-to-network signaling methodology.

2. **Network-to-network Signaling:** The switches in the PSTN intercommunicate with each other on a network-to-network signaling methodology.

User-to-Network Signaling

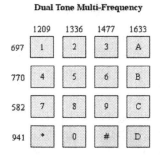

Figure 1.9: Dual Tone Multi-Frequency

Generally the PSTN is connected through analog, Integrated Services Digital Network (ISDN), or through a T1 carrier when using twisted copper pair as the transport. In this type of networking the most common signaling method for user-to-network analog communication is Dual Tone Multi-Frequency (DTMF), an in-band signaling protocol, whereby the tones are transferred through the voice path. *Figure 1.9* shows how DTMF tones are derived. When the numbers are dialed from your telephone handset (as shown in *Figure 1.9*), a tone passes from your phone to the central office switch to which you are connected, and it conveys the request to the switch about the number you want to call.

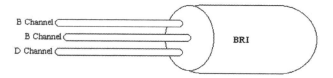

Figure 1.10: Basic Rate Interface

ISDN uses another method of signaling known as out-of-band, which allows transportation on a different channel other than what is used for voice. We already know that the channel on which the voice is carried is called a bearer (or B channel) and is 64 Kbps. Whereas the channel on which the data signal is carried is called a data channel (or D channel) and is 16 Kbps. *Figure 1.10* illustrates a Basic Rate Interface (BRI) that consists of two B channels and one D channel.

There are several benefits associated with Out-of-band signaling, such as:

- Signaling is multiplexed (consolidated) into a common channel.

- Glare is reduced (glare occurs when two people on the same circuit seize opposite ends of that circuit at the same time).

- A lower post dialing delay.

- Additional features, such as higher bandwidth are realized.

- As well, since setup messages are not subject to the same line noise as DTMF tones, call completion is greatly increased.

On the other hand, In-band signaling has a few limitations as well. Such as the possibility for lost tones that occurs when signaling is carried across the voice path. This is a common reason why a user can sometimes experience problems remotely accessing his/her voice mail.

Network-to-Network Signaling

Network-to-network communication is normally carried across the following transmission media:

> ➤ T1/E1 carrier over twisted pair

> > o   T1 is a 1.544-Mbps digital transmission link generally used in North America and Japan.

> > o   E1 is a 2.048-Mbps digital transmission link generally used in Europe.

> ➤ T3/E3, T4 carrier over coaxial cable

> > o   T3 carries 28 T1's or 672 64-Kbps connections and is 44.736 Mbps.

> > o   E3 carries 16 E1's or 512 64-Kbps connections and is 34.368 Mbps.

> > o   T4 handles 168 T1 circuits or 4032 4-Kbps connections and is 274.176 Mbps.

> ➤ T3, T4 carrier over a microwave link

> ➤ Synchronous Optical Network (SONET) across fiber media

> ➤ SONET is generally deployed in OC3, OC12, and OC48 rates, which are 155.52 Mbps, 622.08 Mbps, and 2.488 Gbps, respectively.

Network-to-network signaling types incorporate numerous in-band signaling methods, such as Multi-Frequency (MF) and Robbed Bit Signaling (RBS) that are further used for network signaling methods.

Digital carrier systems (T1, T3) use A and B bits to indicate on/off hook supervision. These A/B bits are designed to follow Single Frequency (SF) tones[4], whereas on the other hand, MF is quite similar to DTMF,

---

[4]   SF typically uses the presence or absence of a signal-to-signal A/B bit transitions

but it utilizes a different set of frequencies. As with DTMF, MF tones are sent in-band. However, instead of signaling from a home to an end office switch, MF signals from switch-to-switch. Network-to-network signaling also uses an out-of-band signaling method known as Signaling System 7 (SS7)[5] (or C7 in European countries). SS7 is a method of sending messages between switches for basic call control and for CLASS. These CLASS services still rely on the end-office switches and the SS7 network. SS7 is also used to connect switches and databases for network-based services (for example, 800-number services and Local Number Portability [LNP]).

Some of the major advantages of moving to an SS7 network are as follows:

➤ Reduced Post-dialing Delay: There is no need to transmit DTMF tones on each hop of the PSTN and experience reduced post-dialing delay. The SS7 network handles the transferring of all the digits in an initial setup message that comprises of the entire calling and called number. When using in-band signaling, each MF tone normally takes 50 ms to transmit. This means there is at least a .5-second post-dialing delay per PSTN hop. This number is based on 11-digit dialing (11 MF tones × 50 ms = 550 ms).

➤ Increased Call Completion: SS7 is a packet-based, out-of-band signaling protocol, compared to the DTMF or MF in-band signaling types. Single packets containing all the necessary information (phone numbers, services, and so on) are transmitted faster than tones generated one at a time across an in-band network.

➤ Connection to the IN: This connection provides new applications and services transparently across multiple

---

[5] SS7 is beneficial because it is an out-of-band signaling method and it interconnects to the Intelligent Network (IN). Connection to the IN enables the PSTN to offer Custom Local Area Signaling Services (CLASS) services.

vendors' switching equipment, as well as the capability to create new services and applications more quickly.

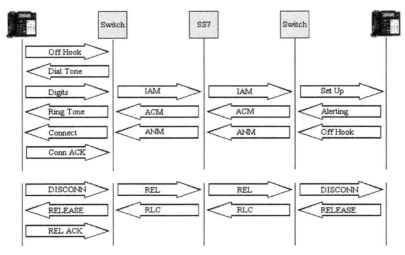

Figure 1.11: PSTN Call Flow to Someone's Home

To further explain the PSTN, let's visualize a call from your house to your friend's house situated 10 miles away. We are considering a situation in which the call traverses an end office switch, the SS7 network (signaling only), and a second end office switch. *Figure 1.11* illustrates the entire network in detail and displays the call flow from your house to your friend's house.

To better explain the diagram in *Figure 1.11*, let's discuss the entire flow of the call step by step:

- You pick up the phone and send an off-hook request to the end office switch.

- The switch sends back a dial tone.

- You dial the digits to call your friend's house. It should be noted that this request is sent in-band through DTMF. The switch reads and analyzes the digits and sends an Initial Address Message (IAM, or setup message) to the SS7 network, which then reads the incoming IAM and sends a new IAM to your friend's switch.

■ Your friend's switch sends a setup message to your friend's phone, which sends a ringing signal to his phone.

■ An alerting message (alerting is the same as the phone ringing) is sent to you're your friend's switch (not from his phone) back to the SS7 network through an Address Complete Message (ACM). The SS7 network reads the incoming ACM and generates an ACM to your switch.

■ At this point, you can hear a ringing sound and know that your friend's phone is ringing. The ringing is unsynchronized; it should be noted that your local switch normally generates the ringing when the ACM is received from the SS7 network.

■ Your friend picks up his phone, sending an off-hook indication to his switch.

■ Your friend's switch sends an ANswer Message (ANM) that is read by the SS7, and a new ANM is generated to your switch.

■ A connect message is sent to your phone (only if it is an ISDN phone) and a connect acknowledgment is sent back (again, only if it is an ISDN phone). If it is not an ISDN phone, then on-hook or off-hook representations signal the end office switch.

■ You can now talk to your friend until you hang up the phone (on-hook indication).

If your friend's phone is busy, you can also use an IN feature by which you can park on his line and have the PSTN call you back after he gets off of the phone.

Now that you have a basic understanding of how the public switched telephone network functions, the next section discusses services and applications that are common in the PSTN.

PSTN Services and Applications

The oldest and hitherto largest telecommunications network in existence is the public switched telephone network that has in excess of 700 million subscribers. For a long time, the PSTN was the only bearer network available for telephony. Today, many people choose the mobile telephone network for their calls. Other bearer networks for voice transmission include integrated service digital network (ISDN), asynchronous transfer mode (ATM), frame relay and the Internet.

Numerous services are now readily available, which were not available just a few years ago. These services come in two common features:

- Custom Calling Features

- CLASS Features

Custom calling features uses an end office switch, not the entire PSTN to transmit information from one circuit-switch to another, whereas, CLASS features require SS7 connectivity to transmit these features from end-to-end in the PSTN.

Some of the popular 'custom calling' features commonly found in the PSTN today are discussed below:

- **Call Waiting**: This feature notifies users who are already engaged with a call that they are receiving another incoming call.

- **Call Forwarding:** This feature enables subscribers to forward incoming calls to a different destination.

- **Three-way Calling:** This feature enables conference calling (up to 3 individual people or parties on the same line).

As discussed earlier, deployment of the SS7 network allows its advanced features to be carried end-to-end. Some of the CLASS features are discussed below:

- **Display:** This feature displays the directory number (caller ID) of the calling party or their Automatic Number Identification (ANI) as used with T1 circuits.

- **Call Blocking**: This feature blocks some specific incoming numbers and sends a message to the caller that their call is not accepted.

- **Calling line ID Blocking:** This feature blocks the outgoing directory number from being shown on somebody else's display.

- **Automatic Callback:** This feature enables users to put a hold on the last number dialed if a busy signal is received, and then place the call after the line is free.

- **Call Return:** This feature enables users to quickly reply to missed calls.

A majority of such features are made possible due to the use of SS7. There are numerous inter-exchange carriers (IXC's) also offering business features, such as:

- **Circuit-switched Long Distance:** Basic long-distance services (normally at a steeply discounted rate).

- **Calling Cards:** Pre-paid and post-paid calling cards. You dial a number, enter a password, and then call your destination.

- **800/888/877/866 Numbers:** The calling party is not charged for the call; rather, the party called is normally charged at a premium rate.

- **Virtual Private Networks (VPN's):** A network that is constructed by using public wires to connect nodes. These systems use encryption and other security mechanisms to ensure that only authorized users can access the network and that the data stream cannot be intercepted.

- **Private Leased Lines:** Private leased lines from 56 Kbps to OC48's enable both data and voice to traverse different networks. The most popular speed by far in North America is a T1.

- **Virtual circuits (Frame Relay or Asynchronous Transfer Mode [ATM]):** The telephone carrier (IXC or LEC) switches your packets. It does this packet-by-packet (or cell by cell in ATM), and not based upon a dedicated circuit.

The above-discussed list of IXC business features is simply a collection of a few examples of the more popular features and applications available within the PSTN. Although technically the oldest network, the PSTN is still in its infancy. While consumers are using more and more of its features, the basic user experience has remained somewhat consistent since the inception of digital networking for telephony communications.

Standards Based Packet Infrastructure Layer

In this new model, the packet infrastructure has replaced the circuit-switching infrastructure. This infrastructure most likely will be IP, even though this model also works if ATM is the underlying transport and IP rides across the top. IP is the basis of this model sinec IP is very user friendly, as the packet infrastructure due to its ubiquitous nature and its de facto application interface. This means that software applications running over IP do not have to be known. IP simply transports the data end-to-end, with no real interest in the payload.

Real-time Transport Protocol (RTP) is utilized in addition to a User Data-gram Protocol (UDP)/IP header to provide time stamping. RTP runs atop UDP and IP and is commonly noted as RTP/UDP/IP. RTP is currently the cornerstone for carrying real-time traffic across IP networks. For example, Microsoft Netmeeting utilizes RTP to carry audio and video communications.

To date, all VoIP signaling protocols utilize RTP, UDP or IP as their transport mechanism for voice traffic. Often, RTP packet flows are known as RTP streams. This nomenclature is used to describe the audio path. In IP networks, it is common and normal for packet loss to occur. As a matter of fact, Transmission Control Protocol/Internet Protocol (TCP/IP) was built to utilize packet loss as a means of controlling the flow of packets. In TCP/IP, it is retransmitted if a packet is lost.

In most real time applications, retransmission of a packet is worse than receiving a bad packet because of the time-sensitive nature of the information. The ITU-T recommends a one-way delay of no more than 150 ms. RTP[6] has a field that stamps the exact time the packet was sent (in relation to the entire RTP stream). This information is known as RTP timestamps and is used by the device terminating/receiving the audio flow. The receiving device uses the RTP timestamps to determine when a packet was expected, whether the packet was in order, and whether it was received when expected. All this information helps the receiving station determine how to tune its own settings to mask any potential network problems such as delay, jitter[7], and packet loss.

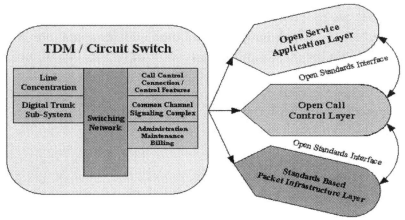

Figure 1.12

---

6   The RTP stream is also often referred to as the media stream. Therefore, you can use IP in conjunction with UDP and RTP to replace a traditional 64-Kbps voice circuit.

7   Jitter is the variation of inter-packet arrival time, or the difference between when a packet is supposed to be received and when it is actually received.

One of the main benefits of IP is the fact that properly built IP networks are self-healing. This means that as dynamic routing protocols are used and multiple possible destinations exist, a network can re-converge based upon the best route. It also means that it is possible for your voice (packetized in IP) to take multiple paths to the same destination. At present, there is a limitation that a user cannot form a single path between two destinations. Each individual packet takes the best path between sender and receiver. The fact that the packet layer is based upon open standards enables multiple vendors to provide solutions that are interoperable. These open standards enable increased competition at this packet layer.

The ITU-T, Internet Engineering Task Force (IETF), European Telecommunication Standards Institute (ETSI), and EIA-TIA are only some of the most popular standards bodies.

One key component of having a standards-based packet infrastructure is the ability to have open standards to layers at the call-control layer. As illustrated in the previous picture - *Figure 1.12*, these open standards are provided by protocols such as H.323, SGCP, MGCP, SIP, and so on, which have open defined interfaces and are widely deployed into the packet infrastructure. One of the jobs of the call-control protocol is to tell the RTP streams where to terminate and where to begin. Call-control accomplishes this task by translating between IP addresses and phone numbering plans.

Open Call-Control Layer

Call-control is defined as a process of making a routing decision on where a call needs to go and then routing the call to the said location. In the PSTN today, these decisions are carried out by SS7 and are made by Service Control Points (SCP's). In this new model of separating the bearers (RTP streams) from the call-control layer and separating the call-control layer from the services, it is necessary to make sure that standards-based protocols are used. Data networking is unique in the fact that multiple protocols can co-exist in a network and you can tailor them to the particular needs of the network.

Many different IP routing protocols exist and each is specifically designed for a certain type of network. Some of the popular ones include the *Router Information Protocol (RIP), Interior Gateway Routing Protocol (IGRP), Enhanced Interior Gateway Routing Protocol (EIGRP), Intermediary System to Intermediary System (IS-IS), Open Shortest Path First (OSPF), and Border Gateway Protocol (BGP).*

Overall the basic functionality of all these protocols is to solve a similar problem i.e. routing updates. Each routing problem is slightly different and requires a different tool. In this case the tool is a routing protocol, which solves each problem. This can be said for all of the VoIP call-control protocols. Besides, they all solve a similar problem, such as phone numbering to IP address translation; however, they might all be used for slightly different purposes.

For instance, H.323 is the most widely deployed VoIP call-control protocol. H.323 however, is not robust enough for PSTN networks, nor advanced IP based telecommunication systems. In order to meet these limitations other protocols such as *Media Gateway Control Protocol (MGCP) and Session Initiation Protocol (SIP)* are being developed.

Research is in the process and it is anticipated that in the coming years, many new VoIP call-control protocols will be developed. Research findings validate that specific requirements are being kept in consideration and each protocol will be developed to fix a certain problem and serve a particular purpose.

VoIP Call-Control Protocols

The main VoIP call-control protocols are *H.323, Simple Gateway Control Protocol (SGCP), Internet Protocol Device Control (IPDC), MGCP, and SIP.*

Discussed below are the basic characteristics of these VoIP protocols:

21

- **H.323** is the ITU-T recommendation with the largest installed base, simply because it has been around the longest and no other protocol choices existed prior to H.323.

- **SGCP** was developed starting in 1998 to reduce the cost of endpoints (gateways) by having the intelligent call-control occur in a centralized platform (or Gateway Controller).

- **IPDC** is very similar to SGCP, but it has many other mechanisms for operations, administration, management, and provisioning (OAM&P) than SGCP. OAM&P is crucial to carrier networks because it covers how they are maintained and deployed.

- In late 1998, the IETF put IPDC and SGCP in a room and out popped **MGCP**. MGCP is basically SGCP with a few additions for OAM&P.

- **SIP** is being developed as a media-based protocol that will enable end devices (endpoints or gateways) to be more intelligent, and enable enhanced services down at the call-control layer.

To briefly explain the various differences between these call-control protocols, let us take a look at how they signal endpoints.

H.323

H.323 is an ITU-T recommended protocol that defines how multimedia traffic is carried over packet networks. H.323 utilizes existing standards (for example, Q.931) to complete this task. H.323 was basically designed to allow multimedia applications to run over unreliable data networks. Voice traffic is only one of the applications for H.323. Most of the initial tasks in this area focused on multimedia applications with video and data sharing requirements. Therefore, H.323 is not the protocol of choice when it comes to robust IP based applications. For example, with respect to H.323, accomplishing a simple call transfer requires a separate specification such as H.450.2. SGCP and MGCP, on the other hand, can accomplish a call transfer with a simple command, known as a modify connection (MDCX), to the gateway or endpoint. This simple example represents the different approaches built into the protocol design itself, one tailored to large deployment for simple applications (MGCP), and the other tailored to more complicated applications but showing limitations in its scalability (H.323). To further demonstrate the complexity of H.323, *Figure 1.13* shows a call-flow between two H.323 endpoints.

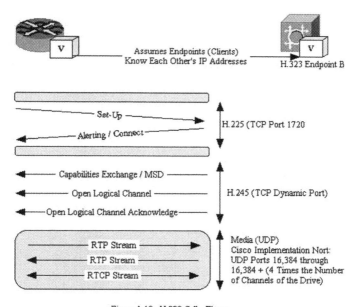

Figure 1.13: H.323 Call - Flows

Figure 1.13 illustrates the most basic H.323 call-flow. In most cases, more steps are needed because gatekeepers are involved.

To better explain *Figure 1.13*, let us step through the call-flow:

1. Endpoint A sends a setup message to Endpoint B on TCP Port 1720.

2. Endpoint B replies to the setup message with an alerting message and a port number to start H.245 negotiation.

3. H.245 negotiation includes codec types like G.729 and G.723.1, port numbers for the RTP streams and notification of other capabilities the endpoints have.

4. Logical channels for the UDP stream are then negotiated, opened, and acknowledged.

5. Voice is then carried over RTP streams.

6. Real Time Transport Control Protocol is used to transmit information about the RTP stream to both endpoints.

This call-flow is based on H.323 v1. However, H.323 v2 enables H.245 negotiation to be tunneled in the H.225 setup message. This is known as fast- start, and it cuts down on the number of roundtrips required to set up an H.323 call.

SGCP and MGCP

SGCP and MGCP were developed to enable a central device known as a Media Gateway Controller (MGC) or soft-switch, to control endpoints or Media Gateways (MG's). Both of those protocols are referenced simultaneously as xGCP. Various applications can be developed through the use of standard-based API's that interface with the MGC's and offer additional functionality, such as call waiting, CLASS features and applications.

The Cisco version of this technology is known as the Virtual Switch Controller (VSC). In this scenario, the entire IP network acts like one large virtual switch, with the VSC controlling all the MG's.

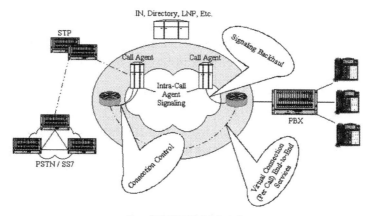

Figure 1.14: Virtual Switch Controller

Figure 1.14 describes how a typical network design works with a virtual switch running MGCP. *Figure 1.14* also illustrates how the legacy PSTN and enterprise networks are connected to gateways or endpoints that enable access into the new packet network. This packet gateway receives direction from the Call Agent (VSC), which can communicate with the SS7 network, the IN, and can tell the gateways or endpoints how and when to set up the call.

To understand *Figure 1.14* in greater detail, let us discuss all the various components involved in this cycle.

- The existing PSTN/SS7 network is connected to the Switching Transfer Point (STP), which also is connected to the MGC or Call Agent. This connection is where the signaling (SS7) takes place.

- The PSTN/SS7 network is also connected to an MG, which is a signal-less trunk that is often known as an Inter-Machine Trunk or IMT. The MG is where the 64-Kbps voice trunks are converted into packets and placed onto the IP network.

- The MGC's or Call Agents also intercommunicate. This protocol is currently undefined in the standards bodies. Based on the current state of the industry, however, it appears that a variant of SIP or ISDN User Part (ISUP) over IP (a portion of SS7 running on top of IP) will be the primary protocol.

The MGC's have a connection to the IN (as described earlier in this chapter) to provide CLASS services. The MGC's receive signals from the SS7 network and tells the MG's when to set up IP connections and with which other MG's they should set them up with.

- The MG on the right side of *Figure 1.14* does not have a connection to the SS7 network. Therefore, a mechanism known as signaling backhaul must be used to tell the VSC when and how a call is arriving. Signaling backhaul is normally done with ISDN. The MG or some other device separates the D channel from the B channels and forwards the D channel to the MGC through IP. Signaling backhaul is currently undefined in the standards bodies.

SIP

SIP is best described by RFC 2543, which states that it is an application-layer control (signaling) protocol for creating, modifying, and terminating sessions with one or more participants. These multimedia sessions include audio, video, and data, and can include multiple partners. SIP enables participants to be invited into an impromptu conference. These multimedia sessions can communicate through multicast, unicast, or a combination of both delivery mechanisms. Very few implementations of SIP are currently running, although many vendors and customers are interested in using SIP to deploy enhanced services.

Open Service Application Layer

When moving to a new infrastructure, it is not necessary to carry over all the features that were on an old infrastructure. Only the features or applications that customers needed are required. Open Service Application Layer is considered to be one of the most interesting layers of any networking protocol. When designing a network that has open interfaces from the packet layer to the call-control layer, and from the call-control layer to the application layer, vendors rather than developing applications, simply write to these standard API's and have access to a whole new infrastructure. When a new packet infrastructure is designed, opportunities for new applications become widely available. Legacy applications such as call-centers for enterprise networks, and standard PSTN applications, such as call waiting and call forwarding, must be ported onto a new infrastructure without the end user realizing that the change has occurred. After these legacy applications are ported, literally thousands of new enhanced applications can be specifically developed for packet infrastructures. These include, but are not limited to; Internet call waiting, push to talk, find me-follow me, and unified messaging.

New PSTN Network Infrastructure Model

As discussed in the last section, the newly created infrastructure can focus on the ability to separate the old stagnant infrastructure into a model by which multiple vendors can develop applications and features quickly for the consumer. *Figure 1.15* shows how Cisco Systems wants to carry this model forward. It describes the relationship between these three layers, as well as the relationship between these layers and the components that would be used in a live network. It is assumed that the carriers will enjoy this method, as it means they will not be locked into a single solution for any of their layers. They will also be able to mix and match all three layers to offer their services, functionality, and the time to market that they need.

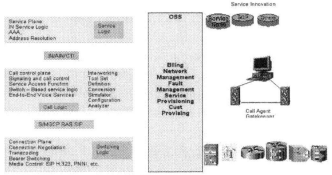

Figure 1.15 *Elements of Packet Telephony*

Amidst all these advantages, there are several carriers who have shown their disinterest in utilizing more than one equipment vendor to cut down on their integration timeframe; however, it is also anticipated that several service providers will also partner with a minimum of two vendors to ensure competition.

Figure 1.15 clearly highlights that the bearers, connection plane, or media transport will be either IP gateways or ATM gateways, or a combination[8] of both. Multiple vendors will be in this space initially, but most likely, they will consolidate to three or five major players. The call-control plane is an extremely important piece of the new PSTN network infrastructure model, as it must gracefully coexist with both the connection plane and the service (application) layers. Many vendors are building MGC technology.

In fact, many researchers are working with approximately 15 vendors to ensure compatibility from the connection plane into the call-control and service/application plane. It is anticipated that numerous vendors will continue to be in the call-control plane, as service providers will more than likely use several vendors for this key technology, depending upon what service they decide to deploy. The responsibility on the Call Agent vendors will be to ensure compatibility from one Call Agent to another. Call Agent interoperability is one of the major mechanism that is instigating service providers in curtailing large-scale, packet-based voice networks. Considering this trend, it is assumed that the service or application plane is where the innovation in the network will happen.

---

[8]   A common trend in the manufacturing and carrier arena is consolidation. The consolidation of manufacturers is one reason for the dramatic reduction in the number of players in this space.

A major concern affecting the service plane[9] is its dependence on soft-switch vendors to open API's that are useful enough to develop services. Due to this, it is perceived that numerous application vendors are taking into account the advancement of Call Agent technology.

Conclusion

Since its inception in 1876, voice and its associated technologies related to the PSTN has become more complex gradually. The PSTN, as you know it today, is on the verge of a revolution. Due to recent technological advancements, the telephone/communications infrastructure is moving to a new model and will soon be able to carry these multimedia conversations. However, the major concern is the requirement of the bandwidth to support these multimedia conversations. This is being accomplished in the bandwidth wars currently being fought by the DSL and cable providers. In the end, consumers will be the ultimate winners, in that they will have access to technology that will eliminate distance barriers, communication barriers, and will truly revolutionize the way things are currently accomplished.

---

[9]   The service plane is where thousands of ISV's converge to develop new and revenue enhancing applications. This same revolution is happening in the PSTN today, and will change the way services and telephony/multimedia networks are designed, built and deployed.

Chapter Two
2) IP Telephony Technology – An Overview

---

a) Emergence of IP Telephony – History Down the Line

    i)   Convergence and IP Telephony

    ii) Evolution of Enterprise Communications from Analog to IP Telephony

    iii) Development of Ethernet Telephones and LAN-Based Telephone Systems

b) Basic Characteristics of Voice versus Data Communications

IP Telephony Technology – An Overview

Definition:

> Internet telephony (IPT) is the transport of telephone calls over the Internet. It does not matter whether they are traditional telephony devices, multimedia PC's or dedicated terminals, which take part in these calls.

Telephony technology, which can be enabled across data networks through Internet protocol, is known as IP Telephony. This is characterized by the set of standards, which enables voice, data and video to coexist over IP based LAN'S, WAN'S, frame relay networks, ATM backbones and the Internet. The applications of which include PC-to-PC communications, PC-to-phone communications and phone-to-phone communications. The convergence of the circuit switched networks such as the Public Switched Telephone Network, or PSTN with packet switched networks such as the Internet, intranet, LAN, WAN and other network types, best explains the nature of IP Telephony, which has created new applications and revenue opportunities thus enhancing the existing communication systems.

With tenure of only five years, the entire market of IP telephony has evolved and has become a lucrative business worth billions, the maxim of which is to provide service to carriers and millions of individual consumers.

Easy money through international toll calls are no more the standard order of the day, as this system reeks havoc on most of the service providers, literally stealing vast profits from toll calls. This is where the IP Telephony concept comes into play and most of the service providers are pressing hard to achieve an IP Telephony solution to maximize their profit and their revenue margin. Here lies the old story whereby the telephone company never looses money; they just change how they make it. Thus explaining the fact that revenues can be achieved from a creative and ever changing array of advanced services, which is mainly through the convergence of voice and data communications, thus pressing on the service providers the need for IP Telephony, which in turn makes the service providers go for the flexible IP Centric OSS to achieve long term success and profitability.

The advantages and the cost effectiveness of IP Telephony is tremendous as it can relieve many of the high costs, skyrocketing global Internet use and the expanding e-commerce, all leading towards a global revolution in the communication sector. Although as of now it is a mere alternative to long distance rates, but the promise that it provides is enormous and can spell wonders in the days to come.

The advantages, or the benefits that may evolve from IP Telephony are summarized below:

- By implementing data oriented switches one could gain access to the switching and multiplexing of data and voice, which may result in better utilization of bandwidth. This can happen by using only a single infrastructure that will serve the purpose for both Internet access and Internet telephony. Like this, it would provide for much lower costs and more profits for clients as well.

- IP Telephony, when implemented is likely to save the high charges of long distance calls, as everything in the future will be linked through the Internet. Statistics show that in Europe the charges of long distance calls are higher as compared to the same in United States.

- The last but not the least important benefit is the routing of software solutions through the Internet. Service providers can extend and integrate software services with other related services, such as white boarding, electronic calendar or WWW, and by implementing true IP Telephony, much lower investments are required as compared to traditional ones.

The fact still remains, that in some cases IP Telephony is not as advanced as it would like to be, and therefore on occasion faces the disadvantages of bad quality voice, and some technical aspects relating to accounting, billing, charging, and roaming, which are yet to be developed fully with IP Telephony.

Structural Design

The IP Telephony system can be better understood if we go into the components of the system. Generally IP Telephony systems are based on the following components:

- End Device, which may be an audio equipped personal computer, single use appliances, a remote IP based telephone, or the most common, a traditional home style standard telephone.

- Gateways, whereby the major task is to translate, encode or decode the messages communicated through the devices, which may happen to be a traditional telephone.

- Gatekeepers or proxies, the function of which is to administer the calls, manage bandwidth, and monitor the authentication of the messages and the monitoring of the translation for the users.

- Multipoint conference units manage or administer the conferences of the parties or the users.

The components as described above can be implemented either as hardware or software and also may be integrated to a single unit and help to communicate with each of the component users over signaling and voice transporting protocols, based on certain standardized codes or languages, so as to ensure smooth functioning of the system.

Emergence of IP Telephony – History down the line

The concept of IP Telephony did not creep up overnight. Contrary to this, the concept of voice communication was in vogue since February 1995 when Vocal Tel first introduced its PC-to-PC Internet phone software. The utility of the so-called voice communication software is to conduct a conversation through their multimedia personal computers.

IP Telephony gateways were introduced in 1996 along with the Telecommunications Reform Act (US). The gateways as described earlier provide a link between the IP networks with that of the traditional PSTN, thus forming the mainstay of the IP Telephony network.

The IP Telephony industry witnessed global expansion during the period 1999 to 2000 with the leading service providers aggressively establishing points of presence in cities all over the world, only to extend their geographical coverage. This resulted in the deployments of affiliate programs focused on establishing relationships with the local service providers in international localities.

The changing trends of the industries came to light only after 2000 and are characterized by:

- The advent of the unified communications resulting in the widespread availability of advanced communications and

- The extension of the IP to local exchanges resulting in full transmission or convergence of voice and data.

The changing trends as mentioned above were enhanced by the following developments that are under process:

- The continued emergence of a new three tiered model in order to replace the traditional circuit switching model and

- The complete integration of the IP Telephony network with that of the traditional SS7 signaling network.

The three-tiered distributed network model has the following structure:

- A standards-based packet infrastructure layer

- An open call control layer

- An open service application layer

The open service application layer, which is comprised of programmable systems enable the service providers to create instantly customer services independent of the hardware vendors. The second development, which is the full development integration of the IP telephony network with that of the traditional SS7 signaling network, enables flawless integration with that of the PSTN network to leverage existing resources and to provide a migration towards full end-to-end IP communications.

Convergence and IP Telephony

The synergy between networks of voice and data has started bringing drastic and far-reaching changes in the field of product development and delivery for Small and Medium size Enterprises (SME's). These current developments are targeted towards crafting one single network framework, which would deliver both data and voice to any type of communication devices effectively and flawlessly. However, at present we are in the Junction of this drive toward synergy of voice and data networks, as the targeted convergence exists only as an idea in principle. There are limited advantages and a slender scope for growth with the current applications, which is on an endeavor of amalgamating the 'traditional circuit switched network' and the 'modern packet-switched networks'. Yet, it is imperative to think about these approaches & techniques that will facilitate these products to develop into solutions that provide significant business advantages.

The recent telecommunications industry trends show that more attention has been paid towards the development of IP Telephony systems and its applications. In general, IP Telephony is simply delivering voice over a network of Internet Protocol. The IP network can be as elaborately surrounding a network backbone of companies like Level 3, AT&T, MCI's, or as distant and minute as the Small Business Enterprise's single-server local area network (LAN). From an end users point of view, an IP telephone system, commonly called an IP system, communications server or LAN PBX, will not only allow voice and data transmission over a single LAN infrastructure, but also radically affect the way of deploying and maintaining voice communication systems, along with changing its way of interacting with their data servers.

The solutions to IP Telephony are fairly new to the industry with the introduction of IP gateways for the Enterprise PBX or key system since the first wave of IP Telephony development, commencing in 1997. An IP-based telephone station on these systems is similar to a digital telephone, the only exception being the packetizing and the sending of voice directly over the LAN through an ethernet connection. Although these early products and its developments were far-reaching and highlighted the potential of IP telephony, they do not actually meet the expectations in terms of regulatory requirements, the standards for reliability, and feature functionality precipitated by the conventional PBX and Key System.

The introduction of a hybrid, IP-enabled digital telephone system is a third alternative for enterprise IP Telephony. This is affected by facilitating a traditional PBX or key system for direct ethernet connection to the LAN. Although the IP-enabled digital telephone system lacks the appeal of IP telephone systems and IP gateways in the trade press, at this point it is proving to be greatly advantageous in the life cycle of these products.

Since the last decade of the 20th century, IP telephony has grown to be an industry altering business opportunity from a hobby for techno geeks. There are numerous vendors that are developing and offering products that are aimed towards making the term "Convergence / Synergy" a reality.

Let's look at the ultimate goal that we are converging upon:

1) Creating a single network infrastructure for replacing contemporary separate packet and circuit switched networks

2) Developing products that effectively and effortlessly carry voice, video and data over a unified network.

The fact is, that IP Telephony is currently the forerunner in this objective.

When voice, video and data communications are simultaneously and jointly delivered over any network that supports the Internet Protocol (IP), it is referred to as IP Telephony. The Internet Protocol is a set standard for software that describes its features as:

1) Keeping track of the Internet work addresses for different nodes

2) Routing outgoing messages

3) Recognizing incoming messages

Principally, it facilitates smooth passage of a data or a voice packet to its final destination via multiple networks. This protocol, by and large is referred to as TCP/IP, and typically works in conjunction with Transmission Control Protocol (TCP).

Like Internet, most LAN's and WAN's use Internet Protocol for data transmission. The poor service quality of the Internet has forced most service providers to build their own IP networks, which are completely controlled by the service provider, carving a much higher service quality. Physical media such as ISDN, xDSL, POTS lines, and twisted pair may support the Internet Protocol too. The foremost benefits of IP Telephony can be summed up as increased managing ability, enhanced collaboration tools, reduced support costs, and better productivity throughout the converged network.

Evolution of Enterprise Communications from Analog to IP Telephony

Since the later half of the 1960's, the pace of evolving technology that powers enterprise communications has accelerated to a considerable extent. The drive to take advantage of the latest network services has served as a fuel for this rapid evolution at the enterprise level. An insight of how the evolution of enterprise communications systems is affected by the introduction of new network services is elaborately dealt with in this section.

Analog in the 1970's

Voice traffic ruled the analog network during the decade of 1970. Voice communications circa 1975 consisted of a Central Office Switch connected to an analog PBX through analog trunks, which in turn was connected to an analog or single line telephone through twisted pair wiring.

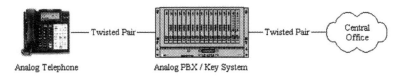

Figure 2.1: Analog Telephone System

1980's: Digital / Circuit Systems, Enterprise LAN

Analog started getting replaced in the early 80's by digital communications due to the introduction of T1 trunks and digital telephones. Digital communications became the standard by the end of 1980. At that time Voice communications was effected by connecting the Central Office switch to the digital PBX using T1 trunks/analog trunks that, in turn, were connected to a digital telephone through twisted pair wiring.

The advent of the desktop PC as a stand-alone device initiated a parallel development in this field during mid-80's. As PC's flourished by the end of 80's, most enterprise' PC's were connected together through the Local Area Network (LAN) that was entirely different from the voice network. LAN's used packet technology instead of following the 'circuit switched technology' based voice network. This resulted in an environment where both LAN's and Voice Networks existed as two entirely different and separate enterprises.

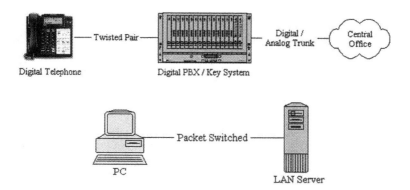

Figure 2.2: Digital Telephone

Early 1990': The Internet Begins to Change Everything

Throughout the decade of the 90's, the enterprise communications paradigm of separate and closed data, and voice networks remained almost unchanged. However, with the spectacular rise of the Internet in 1995 establishing e-mail as a requisite and essential business tool, which subsequently resulted in the proliferation of the Internet Service Provider (ISP), whereby another set of lines was required to connect the Internet with LAN's.

Most enterprises today are using a packet switched network for data, and a circuit-switched network for voice, which remain two separate and outdated communication systems. These individual networks require two sets of hardware, the PBX and the LAN server. As well, two sets of wires to the desktop, two sets of digital lines, one for voice, the other for data, a voice communications provider, an Internet Service Provider, and two sets of management personnel, the LAN Administrator and the Telecom Manager.

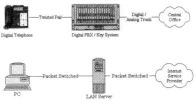

Figure 2.3: Separate Voice and Data Networks

Mid 1990s: The Birth of Computer Telephony and IP Telephony

In the second half of the 90s, computer telephony took birth for the purpose of bridging the gap between the individually wrapped and served data and voice networks. Initially, computer telephony was used to pass call event data, chiefly system status, station status, and call control information from the PBX to the LAN primarily using Computer Supported Telephony Application (CSTA) and Telephony Application Programming Interface (TAPI) that was considered an established set of communications standards. This resulted in the idea of making and answering calls from a PC, the new desktop telephone. It also provided the first opportunity for delivering call events to LAN-based applications in order to enhance customer service and productivity. The birth of "PBX to server communication" and "computer telephony" proved significantly beneficial for businesses in terms of better integration of PBX and server based functions and the ability for Screen synchronization (Pop) Applications.

Late 1990s: Initial Deployment of IP Telephony

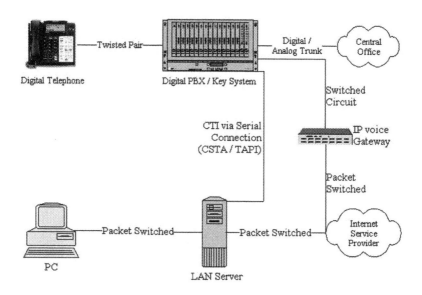

Figure 2.4: IP Gateways Bridge the Gap Between Networks

Early deployment of IP Telephony was for PC-to-PC transmission of voice over the Internet. The transmission of voice between one digital telephone to another through an IP gateway connected to the PBX at both ends was another early example of IP Telephony. Since these early solutions to IP Telephony relied on the Internet, the quality of voice was poor to say the least. Additionally, the lack of standards for Voice over Internet Protocol (VoIP) significantly limited the feasibility of an IP Telephony solution. However, the quality of voice was undistinguishable in effect from that of a conventional long-distance carrier with the use of IP Telephony over a privately managed IP network.

Development of Ethernet Telephones and LAN-Based Telephone Systems

Thanks to the foundation provided by the early developments with IP Gateways and Computer Telephony, a new genre of enterprise telephone systems, that being the IP or packet-based telephone system is rapidly emerging today. Although this new genre of

41

telephone system radically varies in architecture from vendor to vendor, it has a common goal of converging the two different worlds of voice and data communications via a completely packet switched architecture.

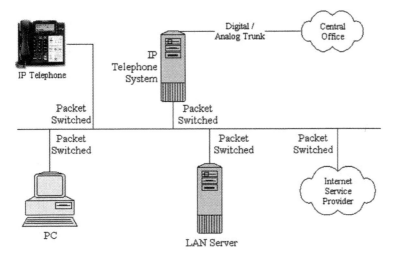

Figure 2.4: IP Telephone System

These systems use IP Telephones, which utilize an Ethernet connection instead of twisted pair cable. The hypothesis is similar to a traditional digital telephone, except that the voice is packetized by the telephone and sent directly to the LAN over a built-in Ethernet connection. Depending upon the architecture, IP Telephones may have the skill of peer-to-peer communications with each other.

In this case, the PBX functionality, which resides inside the LAN-based Telephony Server, is brought into play only when call control is required (e.g. during call transfers, conference calls and voice mail transfers). The end result is a new breed of telephone system where PBX functionality inside an enterprise resides on a LAN-based communications server, and all of the voice traffic flows over the LAN. Gateways are provided to connect directly to the public and private Internet Protocol and circuit switched networks.

The time will certainly come when the LAN-based Telephone System may replace the traditional PBX/Key System. As of now, issues such as limited telephony features, questionable reliability, bandwidth, infrastructure limitations, and regulatory issues prevent wide-scale deployment of LAN-based Telephone Systems.

Fundamental Characteristics; Voice versus Data Communications

Other than a traditional voice network, generally called the public switched telephone network, the sound is transmitted over almost any type of network with the use of voice over data networks. Data networks, such as Asynchronous Transfer Mode backbones, frame-relay networks and the Internet can effectively accommodate voice. Voice can be sent over WAN's and LAN's too by using the TCP/IP protocol. In recent years, nearly all of the attention has been devoted to Internet Protocol, or voice over frame relay.

A strict separation of voice and data traffic is essentially required for traditional network architecture in enterprises and other types of organizations. Earlier, there was no choice other than the one in which low latency was required to support delay sensitive applications like voice and video conferencing, and the data networking didn't deliver the end-to-end Quality of Service. Later on, even after a choice had cropped up, there was an unwillingness to converge the two networks due to various technical, financial and political reasons.

However, converged or unified data and voice networks became more realistic with the continuous increase in the reliability and capability of modern data networks, and the reduction in the cost of bandwidth. Indeed people may raise the concern depending upon the method of analysis that there are compelling cost-related arguments for a converged network.

VoIP - What is Voice over IP?

Voice over IP (VoIP) is the process of converting analog sound into digital packetized data that is then transported over an IP network. Typical implementations requires the involvement of some sort of end-user terminal that can either be a physical appliance looking much like a telephone handset, or an IP phone that can be implemented using software running on a PC with a sound card, speakers and a microphone.

Sound is converted into a data stream for outgoing calls and data into sound for incoming calls by the computer or IP telephone handset (acting as terminal), using most often the PCM codec that is used to encode music on CD's. The handset/virtual handset interfaces with an IP telephony gateway that can either transmit data over your IP network to another station on your network, or to a gateway at a different location, or connect the call to the Public Switched Telephone Network (PSTN). It virtually depends on the location and equipment of the party you are calling.

Key Properties of a VoIP Ready Network

Any VoIP-ready network will have the following four basic characteristics:

- Support for end-to-end Quality of Service (QoS)

- High availability

- Low jitter

- Low latency

- **End-to-End Quality of Service:** The ability to assign different priorities to different types of traffic is described as the Quality of Service (QoS). As an example, assigning a high priority to latency sensitive applications like voice or video would be quite advantageous. Other applications like e-mail can be assigned a lower priority, as they are not affected by the delays. A little latency is affordable for these applications, as long as the information eventually gets through. On the contrary, video and voice traffic are extremely sensitive to delay and disordered packets. The loss or delay of just a few packets can precipitate the disruption and result in an undesirable experience to the user. Thus it is essential to have the ability to assign upper prioritizing to voice traffic over other forms of traffic in order to read an e-mail without cutting-short your phone call.

- **High Availability:** With unified networking, it is critically important to have a reliable and available network. Any network outage will precipitate the loss of all traffic types, be it data, video or voice, resulting not only in the failures control module but also with software upgrades, and reboots, which are transparent to the end user community. Apart from being good to the end-user, it is also a critical feature for the network administrator who can schedule software upgrades at a convenient time rather than a chosen time for securing the user community from the impact. Redundant power supplies, control modules and switch fabrics eliminates single points of failure. Simultaneously, support for VRRP (Virtual Router Redundancy Protocol) that facilitates a way to assign backup routers in a LAN, allows network operators to cluster routers for even greater network availability. These software and hardware features collectively result in what is known as a carrier class platform, which happens to be ideal for carriers, service providers and those running mission critical networks.

- **Low Jitter:** According to the ITU, Jitter is defined as "Short-term variations of the significant instants of a digital signal from their ideal positions in time." Digital communications can be thought of as 1's or 0's sent out on a regular beat. The variation of the timing in the sending of the 1 or 0 off the hypothetical perfect beat would be jitter. Currently available advanced hardware architectures insures low jitter with a non-blocking switch fabric and a design that requires all packets, including those entering and leaving on the same line-card to pass over the back-plane. This increases consistency and reduces jitter, which further optimizes the platform for use in jitter-sensitive applications like IP telephony or multimedia broadcast.

- **Low Latency:** A network device created delay between the receipt of a packet and its retransmission to a different port is termed as latency. Low latency is critical to the end-user's experience in interactive applications. Many consider 250 milliseconds to be the maximum acceptable latency in a VoIP network.

The three most important considerations for implementing a voice-data solution are:

- The perceived voice quality

- The call signaling and call processing requirements

- The data protocol handling and performance attributes of the base data networking products now implementing voice features

Sequentially, the above three aspects are also affected by characteristics of the underlying transport network that may be IP (VoIP), Frame Relay (VoFR), or ATM (VoATM).

Voice Quality is a subjective measure of speech fidelity, which is often rated by a Mean Opinion Score (MOS), and is principally a function of the voice codec and the underlying transport network. 'Voice Codecs' trade-off voice fidelity for efficiencies in bandwidth, latency or computational complexity (processor MIP's). Both the choice of the codec algorithm and the specific implementation of the algorithm affect voice quality. The specific implementation of the algorithm is specifically important in the situation of network congestion and, or error conditions. In order to achieve a high voice quality, 'silence suppression', the use of "comfort noise" in the absence of voice activity and echo cancellation must also be appropriately implemented for the codec.

The other important attribute that affects the preciseness of a voice conversation, besides reliability, is latency (delay). The common consensus of various studies on this subject is that one-way delay of less than 100 milliseconds is undetectable to most users, whereas a delay of 100-200 milliseconds, although perceptible, is tolerable. Delay of more than 200 milliseconds is distinctly annoying, and a delay of over 400 milliseconds results in an effectively half-duplex (speak and wait) conversation. The components of delay include the latency of the transmitting and receiving codecs, and the delay of the underlying transport network, which may include both propagation delay and queuing delays at intermediate routers and switches. In practicality, most of the voice codecs that are currently used operate in the range of 20 to 50 milliseconds of latency, making the network component the dominant contributor of delay.

Public WAN services are now generally offered at several price tiers corresponding to quality of service (QoS) guarantees for performance metrics like latency (delay) and packet loss. For switched services like ATM Service and Frame Relay, guarantees of below 100 milliseconds delay are not rare. Depending on the number of intermediate hops or routers, Internet Protocol (IP) networks characteristically introduce higher amounts of delay. Private Internet Protocol (IP) networks, which implement IP switching will achieve almost the same results as Layer 2 networks (ATM or Frame Relay), whereas delay over the Internet, comparatively, is unmanaged and often exceeds 400 milliseconds.

Some representative guidelines for defining a high and a fair quality voice service in terms of MOS and end-end delay figures are represented in tabular form below (See *Table 1*).

Table 1: Voice Service Definition Guidelines		
	MOS (Kbps)	Delay (MOS)
High Quality Service		
Domestic	3.5 - 4.0	70 - 150 ms
International	3.0 - 3.8	70 - 300 ms
Fair Quality Service		
Domestic	3.2 - 3.8	70 - 200 ms
International	3.0 - 3.5	70 - 400 ms

Last but not the least, it is imperative to comprehend that a variable component of delay called delay variation (or jitter) are also introduced by networks itself. Sophisticated buffer management techniques, which can accommodate moderate amounts of jitter with no impact on voice quality, are implemented by many commercial voice codecs.

However, unbounded delay variation will spill over these buffers, which can result in degraded voice quality and packet loss. Therefore, it is imperative to checkout whether the network service provider is guaranteeing the maximum upper bound of network delay, or just the average amount of delay experienced over a period of time. As a common rule, networks like the Internet that exhibit above-average delay metrics, would most likely exhibit commensurate amounts of delay variation.

The capabilities of the voice-data system to set up, route and tear-down multi-party and point-to-point calls can best be described as 'call signaling' and 'call processing' functions. Such capabilities are especially useful in environments where there are requirements to coexist with and inter-operate with legacy systems and networks like key systems, Private Branch Exchange (PBX) systems, and the Public Switched Telephone Network (PSTN).

Support for PBX signaling protocols like CAS, Q.SIG, ISDN and CCS may be required for that purpose. It may be advantageous, depending on the application, to utilize voice-data systems (e.g. FRAD's /voice routers), the call routing capabilities of the legacy PBX systems, or both. Also, the various standards for voice over Frame Relay, Internet Protocol and ATM define additional requirements and capabilities specific to each environment.

As the voice transmission requires comparatively low bandwidths, data protocols will unquestionably continue to corner the overwhelming share of traffic volume in a converged voice and data network. Consequently, the data protocol handling and performance attributes of converged voice and data products justifiably will continue to be principal, and most important among evaluation criteria, but the profound cost-of-ownership implications will always be the point of discussion.

Chapter Three
Fundamentals of Internet & Data Networking

a) **Data Communication Protocols**

    i) OSI Reference Model

        (1) Application Layer

        (2) Presentation Layer

        (3) Session Layer

        (4) Transport Layer

        (5) Network Layer

        (6) Data Link Layer

        (7) Physical Layer

    ii) EDI (Electronic Data Interchange)

    iii) TCP/IP Model

b) **Primary Network Concepts**

    i) LAN's verses WAN's

    ii) Transmission Channels and Switching Systems

    iii) Layers, Protocols and Interfaces

c) **Voice on Packets (VoIP)**

    i) Configurations – Local and Wide Area

    ii) Quality Issues – Voice Compression, Network Delay, Packet Loss, Echo & Jitter

The concept of computer networking and the present Internet framework traces its route back to the early 1960's, the period in which the entire telecommunication industry was dominated by telephone networks. As we know, any telephony network is based on 'circuit switching', which is used to transfer information from a sender to a receiver. However, due to the growing impact of computer technology, it almost became an inevitability to research various options of integrating computers and telephony technologies together, and further sharing it among geographically distributed users.

This revolution got a considerable boost at the end of the 1970's, when around 200 users were linked to the ARPAnet[10]. January 1, 1983 saw the official deployment of TCP/IP as the new standard host protocol for ARPAnet (replacing the NCP protocol). In the ARPAnet community, many of the final pieces of the present Internet infrastructure were gradually coming into existence. By the time the 1980's were over, the number of users linked to the Internet grew to approximately 100,000. This growth was largely contributed due to various distinct initiatives for creating computer networks linked with numerous universities. This project was undertaken by BITnet (Because It's There NETwork), which also offered e-mail and file transfer facilities within various universities in the Northeast region. Closely on these lines, CSNET (Computer Science NETwork) was formed to associate together with university researchers without access to ARPAnet. By 1986, NSFNET was developed to provide access to NSF-sponsored supercomputing centers. Initiated with a speed of 56 Kbps, NSFNET was handling around 1.5 mbps by the end of 1980's, and was serving as a primary infrastructure linking various regional networks.

In the late 1980's, several other technological advancements led to the development of various important extensions, which allowed

---

[10] ARPANET (Advanced Research Projects Agency Network) It was one of the earliest networks. It provided a vehicle for networking research centers and universities. ARPANET was the basis for the evolution of the Internet.

TCP to implement host-based congestion control. One of the major developments was the origination of the domain name system, which was used to map between a human-readable Internet name (e.g., onestepconnect.com) and its 32-bit IP address.

The 1990's experienced two major events that represented the continued evolution and commercialization of the Web. The first event was ARPANET, which gradually faded. In the 1980's, MILNET and the Defense Data Network had grown to such a level that it was used to handle most of the US Department of Defense' related traffic and NSFNET emerged as the backbone network for connecting regional networks in the United States and national networks overseas. The second major event was the emergence of 'The World' (www.world.std.com) as the first public dialup Internet Service Provider (ISP). In the initial phases of the 1990's i.e. in 1991, NSFNET lifted its limitations on the use of NSFNET for commercial purposes. By the end of 1995, NSFNET was decommissioned and the Internet Service Providers (ISP's) took the responsibility of handling the Internet backbone traffic.

Amidst all these revolutionizing events, the release of the World Wide Web, which brought the Internet into the homes and businesses of millions and millions of people worldwide, became a milestone in the history of evolution in IP Telephony. The Web also served as a platform for facilitating and deploying various applications such as on-line stock trading and banking, streamlined multimedia services, and information retrieval services.

Gradually Internet Protocol was developed and it became a basis for all modern data networks. But many people remain confused by multi-protocol networks. This chapter offers a comprehensive discussion on some of the commonly used data communication protocols.

Data Communications Protocols

Until recent times, there was a discrete distinction between the kinds of systems developed for data transmission over commonly used telecommunication networks and the one developed exclusively for computer-to-computer communications. Increasingly, however, the difference between the Open System Interconnection (OSI) standards developed by the International Standards Organization, the International Telecommunication Union (ITU), and the Internet computer-to-computer communications protocols developed by the Internet Engineering Task Force (IETF) slowly became blurred. As a result, in the present scenario these two systems are seen as integral components of a global information infrastructure.

Numerous forms of digital communications have been standardized in the last century by several authorities that constitute the International Telecommunication Union (ITU). A very good example can be seen in the European region in which the European telecommunications systems are speedily being upgraded to implement the Asynchronous Transfer Mode (ATM) application of the Broadband ISDN (B-ISDN) specification. On a country level, Germany has widely adopted the Integrated Services Digital Network (ISDN) standard as the basic infrastructure for connecting offices and homes to local exchanges. However, in other countries, it is mandatory to route all digital communications as analog signals through twisted copper-wire pairs that form the backbone of existing analog telephone systems. In such networks, there is a limitation that before the transferring of these digital signals, they need to be converted into analog form using a signal modulator/demodulator (modem).

In order to provide the facility of transferring data up to 56 Kbps using the latest set of standards, data compression techniques are increasingly being integrated with standards for the modulation of digital signals, which hardly matches the speed of a single 64k ISDN channel. It should be kept in mind that ISDN signals can be linked across 6 channels to support broadband transmission services.

There are several public organizations that have created 'Data Communication standards'. Some of the popular ones are:

- **IETF:** Internet Engineering Task Force

- **ISO / IEC JTC1/SC6:** JTC1 is the first (and only) Joint Technical Committee of ISO and IEC, and deals with Information Technology. SC6 is the subcommittee of JTC1, which deals with Telecommunications and Information Exchange between Systems.

- **ISO / IEC JTC1 / SC32:** SC32 is the subcommittee of JTC1, which deals with Data management and interchange.

- **ITU:** International Telecommunication Union (formerly CCITT: Comité Consultatif Internationale de Téléphones et Télégraphes).

- **ETSI:** The European Telecommunications Standards Institute has two sub-committees within the Terminal Equipment (TE) Technical Committee working on data transfer aspects of OSI communications:

  o TE3, the Message Handing Systems sub-committee

  o TE6, the Directory Systems sub-committee

All these organizations work in a group and act as a single team of organizations, whereby, when an organization develops any standard, all other organizations work together to enhance that specific standard.

OSI Reference Model

In the infant days of computing, there was a crowd of different computer manufacturers who developed widely diverse hardware, operating systems and application software. The diverse strengths and weaknesses of individual computer types made them more suitable to some applications (i.e., uses) than others. Thus, enterprise consumers started 'collecting' hardware from a variety of manufacturers for their specific departmental functions (e.g., for bookkeeping, personnel records, order-taking, stock-keeping, etc.). The business efficiency profit in computer systems of various departments rapidly justified the individual investments and also resulted in quick economic payback. However, the demands on computers and computer manufacturers gradually decreased, as the IT (information technology) departments of various organizations started looking for various ways to interconnect their numerous systems, rather than have to manually re-type output from one computer to become input for another. Resultantly, this created a demand to develop a standard means for representing computer information (called data) so that it could become a universal standard, which any computer can understand. On similar lines, there was a need for a standard means of electronic conveyance between systems. To meet these demands, the first universal standard was introduced, which we now refer to as *Open Systems Interconnection (OSI)* standards.

Let's start with the discussion on the layered functions, which make up the OSI (Open Systems Interconnection) model.

Computers and data networks are complex, as they cannot think like humans. These networks have major limitations, in that computers have no 'common sense' unless we program it into them, and thus they need programming in advance to react to any situation. Hence, programmers have to ascertain that every situation that might possibly arise has to have been thought about and a suitable action must be programmed into it in advance.

Now if we are setting up a network for data transmission, it is very important that computers communicate or talk with each other. In such a scenario, if computer A attempts to talk to computer B, it needs to check that computer B is hearing. It needs to check that it is talking to the right piece of equipment within computer B. More so, to communicate these computers also need a common language and must use an agreed set of alphabetic characters.

The OSI Model

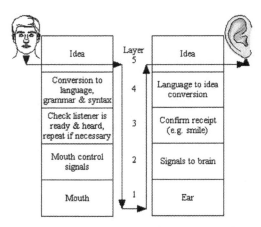

Figure 3.1: A layered protocol model for simple conversation

Initially launched in 1983 by the International Organization for Standardization (ISO), the Open Systems Interconnection (OSI) model categorized the various data communications functions into seven interacting (but independent) layers. The basic idea is to create a modular infrastructure, allowing different standard functions to be jointly combined in a flexible manner to allow any two systems to communicate with one another.

To understand the OSI model let us start with an analogy derived from a simple exchange of ideas in the form of a dialogue between two people. The speaker has to translate his thoughts into words. The words are then converted into sound by nerve signals and appropriate muscular responses in the mouth and throat. The listener in the meantime is busy converting the sound back into the original idea. While this is going on the speaker needs to make sure in one way or another that the listener has received the information and has understood it. If there is a lapse in any of these activities there is no assurance that the original thought would be correctly conveyed between the two parties.

It should be noted that in this example each activity is independent of every other activity. The idea starts at the top of the talker's stack of functions and is converted by each function in the stack until at the bottom it turns up in a sound-wave form. A reverse conversion stack, used by the listener, re-converts the sound waves back into the idea. Each function in the protocol stack of the speaker has an exactly corresponding, or so-called peer function in the protocol stack of the listener. The functions at the same layer in the two stacks correspond to such an extent, that if we could conduct a direct peer-to-peer interaction, we would actually be unaware of how the functions at the lower layer protocols had been undertaken.

For example, let us replace layers 1 and 2 by using a telex machine instead, which by the way was the very first product that I ever sold in this industry – that being the 3M Whisper Telex. For you readers below the age of 30, this is what we used prior to facsimile machines. Now the speaker still needs to think up the idea, correct the grammar and see to the language translation. Although, instead of being aimed at mouth muscles and sound waves, finger muscles and telex equipment do the rest, provided the listener also possess' a telex machine. We cannot, however, simply replace only the speaker's layer-1 function (the mouth), if we do not carry out simultaneous peer protocol changes on the listener's side, because an ear cannot pick up a telex message.

As long as the layers interact in a peer-to-peer manner, and as long as the interface between the function of one layer and its immediate higher and lower layers is unaffected, it is insignificant how the function of that individual protocol layer is carried out. This is the principle of the Open Systems Interconnection (OSI) model and all layered data communications protocol.

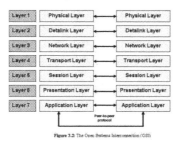

Figure 3.2: The Open Systems Interconnection (OSI)

The OSI model sub-divides the function of data communication into seven layered and peer-to-peer sub-functions, as shown in *Figure 3.2*. Respectively from layer 7 to layer 1 these are called; the application layer, presentation layer, session layer, transport layer, network layer, data link layer and the physical layer. Each layer of the OSI model relies upon the service of the layer beneath it. Thus the transport layer i.e. layer 4 relies upon the network service, which is provided by the stack of layers 1 to 3 beneath it. Similarly the transport layer provides a transport service to the session layer, and so on. The functions of the individual layers of the OSI model are defined more fully in ISO standards (ISO 7498), and in ITU-T's X.200 series of recommendations.

In a nutshell, they are as follows:

The Application Layer (Layer 7)

The Application Layer provides communication function services to match all conceivable types of data transfer, control signals and responses between the linked computers. A wide range of application layer protocols has been defined to accommodate all sorts of different computer equipment types, activities, controls and other applications. These are usually defined in a modular fashion, the simplest common functions being termed application service elements (ASE's), which are sometimes grouped in specific functional combinations to form application entities (AE's) – standardized communications functions, which may be directly integrated into computer programs. These communications functions or protocols have the appearance of computer programming commands (e.g., get, put, open, close, etc.). The protocol sets out how the command or action can be invoked by a given computer program/application and the sequence of actions, which will result in the peer computer (i.e., the computer at the remote end of the communication link). By standardizing the protocol, we allow computers to talk and control one another without misuse or misinterpretation of requests or commands.

The Presentation Layer (Layer 6)

The presentation layer is responsible for ensuring that the data format of the application layer command is suitable for the recipient. The presentation layer protocol informs the recipient about the language, syntax and character set of the application layer command (in other words, which particular application layer protocol is in use).

The Session Layer (Layer 5)

A data communication session between two computers is almost similar to a conversation session between two people. Thus, in a communication session, the two devices at each end of the communication medium must conduct their conversation in an orderly and timely manner. They must listen when spoken to, repeat as necessary, and answer the questions properly. The session protocol regulates the conversation and thus includes commands such as start, suspend, resume and finish, but does not include the actual content of the communication.

The session protocol is rather like a tennis umpire. He/she cannot always tell how hard the ball has been hit, or whether there is any spin on it, but he/she knows who has to hit the ball next and whose turn it is to serve, and he/she can advise on the rules when there is an error in order that the game may continue. The session protocol negotiates for an appropriate type of session to meet the communication need, and then it manages the session.

A session may be established between any two computer-applications, which need to communicate with one another. In this sense the application may be a window on the computer screen, or an action or process being undertaken by a computer. Since more than one window may be active at a time, or more than one task may be running on the computer, it may be that multiple windows and tasks are intercommunicating with one another by means of different sessions. During such times, it is important that the various communication sessions are not confused with one another, since all of them may be sharing the same communications medium (i.e., all may be taking place on the same 'line').

As discussed above, Open System Interconnection (OSI) data communication standards belong to the Application Layer (i.e. Layer 7) of the OSI Basic Reference Model (ISO 7498:1994). The Application Layer is specified in terms of "application contexts" and using building blocks called "Application Service Elements" (ASE's). It resides above the Presentation Layer (i.e. Layer 6), which identifies alternative encodings, and the Session Layer (i.e. Layer 5), which provides dialogue control. Collectively, the three layers provide Application Services, and are commonly referred to as the *Upper Layers.*

The *Lower Layers* of the OSI stack are Transport (Layer 4), Network (Layer 3), Link (Layer 2) and Physical (Layer 1).

The Transport Layer (Layer 4)

The transport layer provides the end-to-end data relaying service needed for a communication session. This layer is responsible for establishing a transport connection between the two end-user devices (e.g., Windows or Tasks) by selecting and setting up the best network connection to match the session requirements in terms of destination, quality of service, data unit size, flow control, and error correction needs. If more than one network is available (e.g., lease-line, packet-switched network, telephone network, router network, etc.), then the transport layer chooses between them.

One of the most remarkable features of the transport protocol is its ability to set up reliable connections in cases even when multiple networks need to be traversed in succession (e.g., a connection travels from LAN (Local Area Network) via a Wide Area Network to a second LAN). The IP-related protocol transmission control protocol (TCP) is an example of a transport layer protocol, and it is this single capability of TCP combined with IP (TCP/IP), that has made the IP-suite of protocols so widely used and accepted.

Besides this, the transport layer is also responsible for supplying the network addresses needed by the network layer for the correct delivery of a message. The network address may be unknown by the computer application using the connection. The mapping function provided by the transport layer in converting transport addresses (provided by the session layer to identify the destination) into network recognizable addresses (e.g., telephone numbers) shows how independent the separate layers can be: the conveyance medium could be changed and the session, presentation and application protocols could be quite unaware of it.

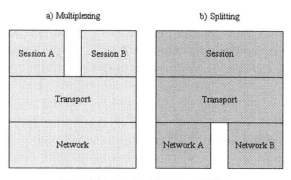

Figure 3.3 Protocols Multiplexing and Splitting

The transport protocol is also responsible for some important multiplexing and splitting functions (*Figure 3.3*). In its multiplexing mode the transport protocol is capable of supporting a number of different sessions over the same connection, rather like playing two games of tennis on the same court. Humans would get confused about which ball to play, but the transport protocol makes sure that computers do nothing of the kind.

Two sessions from a mainframe computer to a PC in a remote branch site of a large shopping chain might be used simultaneously to control the building security system and (separately) to communicate customer sales. Different software programs (for Security and for Sales) in both the mainframe computer and in the PC could thus share the same telecommunications line without confusion. Conversely, the splitting capability of the transport protocol theoretically allows one session to be conducted over a number of parallel network communication paths, like getting different people to transport the various volumes of an encyclopedia from one place to another.

The transport protocol also caters for the end-to-end conveyance, segmenting or concatenating (stringing together) the data, as the network requires.

The Network Layer (Layer 3)

The network layer sets up and manages an end-to-end connection across a single real network, determining which permutation of individual links need be used and ensuring the correct transfer of information across the single network (e.g., Local Area Network or Wide Area Network). Examples of layer-3 network protocols are IP and X.25.

The Data Link Layer (Layer 2)

The data-link layer functions on an individual link or sub-network part of a connection, handling the transmission of the data across a particular physical connection or sub-network (e.g., LAN - Local Area Network) so that the individual bits are transmitted over that link without error. ISO's standard data-link protocol, specified in ISO 3309, is called high level data link control (HDLC). Its functions are to:

1. Synchronize the transmitter and receiver (i.e., the link end devices);

2. Control the flow of data bits;

3. Detect and correct errors of data caused in transmission;

4. Enable multiplexing of several logical channels over the same physical connection.

Typical commands used in data-link control protocols are thus ACK (ACKnowledge), EOT (End Of Transmission), etc. Another example of a 'link' protocol is the IEEE 802.2 logical link control (LLC) protocol used in Ethernet and Token Ring LAN's.

The Physical Layer (Layer 1)

The physical layer is concerned with the medium itself. It defines the precise electrical, interface and other aspects specific to the particular communications medium.

Examples of physical media in this layer are:

- The cable of a DTE/DCE interface as defined by EIA RS-232 or ITU-T recommendations: X.21, V.35, V.36 or X.21bis (V.24/V.28);

- A 10 Mbit/s Ethernet LAN based on twisted pair (the so-called 10baseT medium);

- A 4 Mbit/s or 16 Mbit/s Token ring LAN using Twinax (i.e., 2 x coaxial cable);

- A digital leased line (e.g., conforming to ITU-T recommendation I.430 or G.703);

- A high-speed digital connection conforming to one of the SONET (Synchronous Optical NETwork) or SDH (synchronous digital hierarchy) standards (e.g., STM-1, STM-4, STM-16, OC3, OC12, STS3 etc.);

- A fiber optic cable;

- A radio link.

EDI (Electronic Data Interchange)

By the 1980s, organizations were able to interconnect their different department computer systems together for book-keeping, order-taking, salaries, personnel, etc., and the focus of development shifted towards sharing computer data directly with both suppliers and customers. They started to think; *"Why take an order over the telephone when the customer can submit it directly by computer, eliminating both the effort of taking down the order and the possibility of making a mistake in doing so?"* This concept instigated large retail organizations and the car manufacturers to jump on the bandwagon of EDI (Electronic Data Interchange).

Electronic Data Interchange (EDI) is the transmission between businesses of information in a standard computer-readable format. It includes electronic order placement, electronic shipping notification, electronic invoicing, and many other business transactions that computers can actually perform better than people.

Many business documents can be exchanged using EDI, but the two most common are purchase orders and invoices. At a minimum, EDI replaces the mail preparation and handling associated with traditional business communications. However, the real power of EDI is that it standardizes the information communication communicated in business documents, which makes possible a paperless exchange. The application of EDI in trade transactions enables the transfer of data from one computer to another in such a way that each transaction in the trading cycle (for example, commencing from the receipt of an order from an overseas buyer, through the preparation and lodgment of export and other official documents, leading eventually to the shipment of the goods) can be processed with virtually no paperwork.

The challenge of Electronic Data Interchange (EDI) between different organizations is considerably greater than the difficulties of 'mere' interconnection of different computers as originally addressed by OSI. When data is transferred only from one machine to another within the same organization, then that organization may decide in isolation, which information should be transferred, in which format and how the information should be interpreted by the receiving machine. But when data is moved from one organization to another, at least three more problems arise in addition to those of the interconnection:

- The content and meaning of the various information fields transferred must be standardized (e.g., order number format and length, address fields, name fields, product codes and names, etc.).

- There needs to be a means of reliable transfer from one computer to the other, which allows the sending computer to send its information independently of whether the receiving computer is currently ready to receive it. In other words, the 'network' needs to cater for store-and-retrieve communication between computers (comparable with having a postbox at the post office for incoming mail, which allows you to pick up your mail at a time convenient to you as the receiver).

- There needs to be a way of confirming correct receipt.

Various new standardization initiatives emerged to support EDI, among the first of which were:

- The standardization of bar codes and unique product identification codes for a wide range of grocery and other retail products was undertaken. The industry-wide standard codes provided the basis for the 'just-in-time' re-stocking of supermarket and retail outlet shelves on an almost daily basis by means of EDI, which has helped in sky rocketing companies like Wall Mart to the top of the food chain.

- The major car manufacturers demanded EDI capability from their component suppliers, so that they could benefit from lower stock levels and the associated cost benefits of JIT (just-in-time) ordering. Car products and components became standardized too.

- The banking industry developed Electronic Funds Transfer at the Point-of-Sale (EFTPOS) for ensuring that your credit card could be directly debited while you stood at the register.

All of the above are examples of EDI, and whole data networking companies emerged, specializing in the needs of a particular industry sector, with a secure network serving the particular community of interest. Thus, for example, the ODETTE network provided for EDI between European car manufacturers. TRADERNET was the EDI network for UK retailers. SWIFT is the clearing network of the banks, and SITA was the network organization set up as a cooperative venture of the airlines for ticket reservations and flight operations. Subsequently, some of these networks and companies have been subsumed into other organizations, but they were important steps along the road to modern electronic business. The store-and-retrieve methods used for EDI include e-mail and the ITU's message handling system (MHS). Both are application layer protocols, which cater to the store-and-retrieve method of information transport, as well as the confirmation of reply.

EDI works on any type of computer hardware, application system, or business process used. The people using EDI are called Trading Partners. Two of the most important things required to implement an EDI system are EDI software, and the means for electronic communications. EDI software has the following characteristics and benefits to meet various business requirements.

- **Reduce administrative expenses:** Document handling, mailing, and document management expenses are reduced, as everything can be done electronically.

- **Time Saving:** Via electronic communications, information reaches trading partners faster.

- **Improve internal management:** Clerical workloads are lessened, phone calls are fewer, and there is less chance of errors being made resulting in timely production or reports.

- **Improve trading relationships**: Communication delays are eliminated, thus the transfer of business information (orders and sales) is accomplished quickly, resulting in better relationships with trading partners.

- **A competitive advantage over non-EDI users**: More companies choose to deal with suppliers who use EDI, which gives them a bigger share of the market than non-EDI users.

- **Easy to upgrade:** EDI trading is constantly advancing; therefore it is easy to upgrade in response to any change and enhancement.

- **Multi-network connectivity**: It must not restrict connections to major EDI networks (VAN's).

- **Multi-standard capability**: Nowadays, trading partners need to exchange electronic documents of different standards. The software must accommodate this need.

- **Allows printing**: Some companies need a hard copy of incoming messages. This facility should be included in the software.

- **Ease of mapping:** It should provide user-friendly mapping processes, which are a necessity to large companies with many trading partners.

As seen in the diagram, EDI follows these steps:

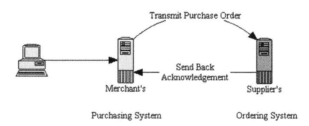

Transmit Purchase Order

Send Back
Acknowledgement

Merchant's   Supplier's

Purchasing System   Ordering System

- The EDI Translation software of the sender translates the document into a standard format and establishes the communications connection, usually by dialing the telephone number of the VAN.

- The message from the computer file of the sender will be transferred to an electronic mailbox on the VAN.

- From the electronic mailbox, the trading partner's software will retrieve the file, interpret the message, check that it complies with the EDI standards and store it. It will also send a Functional Acknowledge to inform the sender if the message was received and complied with the EDI standard.

• To read and process the interpreted message, the receiver has two options. Either use the EDI Translation software to produce a hard copy of the message or to restructure (map) the message into a format required by the existing computer application before further processing.

TCP/IP Model

TCP/IP generally refers to network architecture and protocols that are closely bound: a transport protocol, TCP (Transmission Control Protocol), IP (Internet Protocol). TCP/IP model is network architecture of 4 layers in which the TCP and IP protocols have a major role, as they comprise the official and most common implementation. From a broader perspective, TCP/IP means two different things: the 4-layer network architecture and the set of 2 protocols, TCP and IP.

Today the TCP/IP model has gradually become the universal model of reference and is replacing the OSI model. If we see the origin of TCP/IP we find that it dates back from the ARPANET network[11]. When TCP and IP protocols were invented in 1974, the ARPA signed several agreements with manufacturers (especially BBN) and the Berkeley University, where a Unix system was under development to impose this standard, which was done.

In the present scenario, there is always a great deal of discussion about the comparative advantages and disadvantages between the OSI standards and Internet protocols. In this context, it is very important to note that there is no direct similarity between the OSI application standards and the Internet application level protocols. Differences

---

[11] ARPANET is a telecommunication network developed by the ARPA (Advanced Research Projects Agency), the research agency of the American ministry of defense (the DOD: Department Of Defense).

in their technical nature are closely related to the very different fundamental principles that underlie the OSI Reference Model and the Internet protocol suite, and the very different manners in which OSI standards and Internet protocols have been developed:

- OSI application standards are based on upper layer architecture and discrete modules. The three upper layer stacks provide full inter-working flexibility (some would say at a cost of complexity). The OSI upper layers are in principle independent of the OSI Transport Layer and Network Layer.

- Internet application protocols are written from the ground up. TCP/IP based application services such as FTP and TELNET are plugged directly into the Transport services

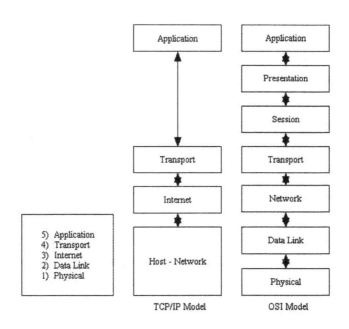

The TCP/IP model does not exactly match the OSI model. There is no universal agreement regarding how to describe TCP/IP with a layered model but it is generally agreed that there are fewer levels than the seven layers of the OSI model. The TCP/IP model superseded the OSI model. This is the model that is currently most widely used. The various layers in the TCP/IP model are:

Note: The Sessions and Application Layer present in the OSI Model are absent in the TCP/IP Model.

These differences have also assisted in fueling the technological advancements within the industry. Deliberations over the co-existence and interoperability of OSI, TCP/IP and proprietary protocol suites in recent years have resulted in a number of specifications on multi-protocol operation, which aim to provide an Any-to-Any solution (any application service over any lower layer protocol stack).

Comparison with the OSI Model

Let's start with the basic similarity between OSI and TCP/IP models.

The TCP/IP and OSI models are both based on the concept of independent protocol stacks. Then, taken as a whole, functions are the same.

Now if we move on to the basic differences, we find that the OSI model makes clearly the difference between 3 main concepts, although TCP/IP does not make this distinction. These 3 concepts are services, interfaces and protocol. Indeed, TCP/IP does not make a clear difference between these concepts, in spite of the efforts of designers to bring the TCP/IP model closer to the OSI model. This is due to the fact that the TCP/IP model protocols appear first, before specifications. The model is finally a theoretical justification of protocols, without making them independent to each other.

Finally, the last important difference comes from the mode of connection. Indeed, connection-oriented modes and connectionless modes are available in both models, but not at the same layer. In the OSI model, these modes are only available at the network layer (at the transport layer, only the connection-oriented mode is available), though they are available at the transport layer in the TCP/IP model (the Internet layer only offers the connectionless mode). Therefore, TCP/IP has an advantage compared to the OSI model as applications, which directly use the transport layer have the choice between both modes.

There are a number of reasons for the success of the TCP/IP model over the OSI model:

1) Internet is built on the foundation of the TCP/IP suite. The tentacles of the Internet and the World Wide Web have spread throughout the world and that is fundamentally the main reason for the success of the TCP/IP model over the OSI model.

2) TCP/IP protocols were initially researched under a project in the Department of Defense (DOD). DOD was committed to international standards and the OSI model could not meet most of its operational requirements. So it started to develop TCP/IP. Since the DOD is the largest consumer of software products in the world, the vendors were encouraged to develop TCP/IP based products.

The layers of the TCP/IP model are defined as follows:

Application Layer

In TCP/IP, the Application Layer also includes the OSI Presentation Layer and Session Layer. The application layer determines the presentation of the data and controlling of the session. In TCP/IP the terms socket and port are used to describe the path over which applications communicate. There are numerous application level protocols in TCP/IP, including Simple Mail Transfer Protocol (SMTP) and Post Office Protocol (POP) used for e-mail, Hyper Text Transfer Protocol (HTTP) used for the World-Wide-Web, and File Transfer Protocol (FTP). Most application level protocols are associated with one or more port number.

Transport Layer

In TCP/IP, there are two Transport Layer protocols. The Transmission Control Protocol (TCP) guarantees that information is received as it was sent. The User Data-gram Protocol (UDP) performs no end-to-end reliability checks.

Internet Layer

The Internet Protocol (IP) is normally described as the TCP/IP Network Layer, and because of the Inter-Networking emphasis of TCP/IP, this is commonly referred to as the Internet Layer. All upper and lower layer communications travel through IP as they are passed through the TCP/IP protocol stack.

Network Access Layer

In TCP/IP, the Data Link Layer and Physical Layer are normally grouped together. TCP/IP makes use of the existing Data Link and Physical Layer standards rather than defining its own. Most RFC's that refer to the Data Link Layer describe how IP utilizes the existing data link protocols, such as Ethernet, Token Ring, FDDI, HSSI, and ATM. The Physical Layer typically defines the characteristics of the hardware that carries the communication signal. This describes attributes such as pin configurations, voltage levels, and cable requirements. A few examples of Physical Layer standards are RS-232C, V.35, and IEEE 802.3.

The four-layer structure of TCP/IP is built as information is passed down from applications to the physical network layer. When data is sent each layer treats all of the information it receives from the layer above as data, and attaches control information to the front of that data. This control information is called a header, and the addition of a header is called encapsulation. When data is received, the opposite procedure takes place as each layer removes its header before passing the data to the layer above.

As shown in the following illustration, each layer of the TCP/IP model corresponds to one or more layers of the seven-layer Open Systems Interconnection (OSI) reference model proposed by the International Standards Organization (ISO).

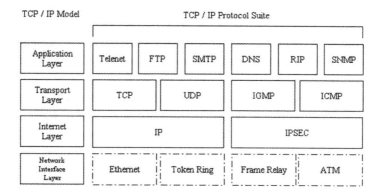

The types of services performed and protocols used at each layer within the TCP/IP model are described in more detail in the following table:

Layer	Description	Protocols
Application	Defines TCP/IP application protocols and how host programs interface with transport layer services to use the network.	HTTP, Telnet, FTP, TFTP, SNMP, DNS, SMTP, X Windows, other application protocols
Transport	Provides communication session management between host computers. Defines the level of service and status of the connection used when transporting data.	TCP, UDP, RTP
Internet	Packages data into IP data-grams, which contain source and destination address information that is used to forward the data-grams between hosts and across networks. Performs routing of IP data-grams.	IP, ICMP, ARP, RARP
Network interface	Specifies details of how data is physically sent through the network, including how hardware devices that interfaces directly with a network medium, such as coaxial cable, optical fiber, or twisted-pair copper wire electrically signal bits.	Ethernet, Token Ring, FDDI, X.25, Frame Relay, RS-232, v.35

In addition to multi-point configurations like Ethernet and Token Ring networks, TCP/IP can be configured to operate across serial, point-to-point communication links, such as direct serial cables, dial-up POTS modems, and ISDN lines. Managing communication across this type of link requires different techniques than those used on multi-point networks, because these types of media are often slower and prone to errors in transmission. If the connection travels across a switched network, such as the telephone system, the protocols must also deal with feature negotiation and user authentication each time a link is established, to verify whom the connection is to and how to conduct the conversation.

There are two serial line protocols that are commonly used with TCP/IP namely SLIP and PPP. Both of these encapsulate IP data-grams and transmit them across serial communications lines. PPP is considerably more robust that SLIP.

SLIP

Serial Line Internet Protocol (SLIP) is a Network Access Layer protocol designed for point-to-point serial connections running TCP/IP. It is an older protocol originally designed for connecting two computer systems across a telephone modem line or direct cable connection. It is used most often across dedicated phone lines and sometimes for dial-up. The major use of SLIP is connecting two sites that are using different hardware or software and which share no other common Network Access Layer protocols. It is defined in RFC 1055 and is implemented on most UNIX platforms. Microsoft Windows Dial-Up Networking also supports dial-up connections to a SLIP host, and SLIP support is available on most other hardware and software platforms.

The greatest advantage of SLIP is that it is a fairly easy protocol to implement, and so it is widely available. The greatest drawback of SLIP is that it is a simple-minded application that provides no error detection or correction, compression, or any other advanced features, leaving that to other protocol layers. In most applications, SLIP has been replaced by PPP (Point-to-Point Protocol).

PPP

Point-to-Point Protocol is a Network Access Layer protocol designed for reliable point-to-point serial connections. PPP can be used to encapsulate and transmit IP data-grams, and can also be used to support other protocols (e.g. DECNet, NetBEUI, etc.) across the same connection at the same time. PPP is the most commonly used protocol for POTS dial-up connections to the Internet. It allows a central server to accept incoming calls from many different types of remote clients and negotiates a common set of features for each connection. PPP can also be used to establish permanent connections across a serial line. Most Point-to-Point Protocol dial-up connections can be configured to request a specific IP address, and to re-dial and re-establish a connection automatically if the line is dropped. It is defined in RFC 1331, published in 1992, and is a widely available protocol supported by almost all hardware and software platforms that support TCP/IP and serial connections.

Though more commonly found in single-user applications, Point-to-Point Protocol could also be used to support network dial-in connections. When properly configured, IP for an entire network can be routed across a PPP connection. This way an entire local area network can be connected to a remote network using a single POTS, ISDN, or direct serial line. Multiple systems on the local network can be assigned IP addresses and given access to the remote network. If the remote network is an Internet Service Provider, the local network users can be given access to the Internet across the PPP line.

PPP uses a number of protocols to manage the connection between two systems. The PPP Link Control Protocol (LCP) is used to establish and test the data link between the two systems at either end of a communications line. LCP is able to negotiate a rich assortment of configuration parameters and provides additional management functions.

Two authentication protocols namely Password Authentication Protocol (PAP), and Challenge Handshake Authentication Protocol (CHAP) are normally used with PPP. These protocols are documented in RFC 1334 and RFC 1994. Authentication is not a mandatory step, but it can be specified during LCP negotiation. The authentication step normally occurs immediately following LCP completion. PAP and CHAP are intended for use by systems that connect to a PPP server via dial-up lines or switched circuits, but might be used on dedicated links as well. The PPP server can use the identification of the connecting system to select options during network layer negotiation.

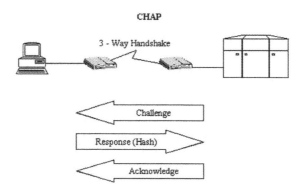

PAP is a simplistic authentication method that does not use strong security techniques. It passes an ID and Password, in clear text, from the client to the server for verification. The client repeatedly passes the ID and Password pair until an acknowledgment is received or the connection is terminated. The passing of the ID and Password in clear text is not considered a secure method because it is possible to eavesdrop on a connection and capture this information.

CHAP is a much more robust authentication method that uses a 3-way handshake. The handshake takes place immediately after LCP negotiation is complete, and may be repeated any time after the link has been established. In the 3-way handshake, the server performing the authentication sends a 'Challenge' message to other system. The challenged system responds with a value that has been calculated using a One-way Hash function based on data from the message and an ID and Password pair. The challenging system checks the response against its own calculation of the expected hash value. If the values match then a message is sent to acknowledge the authentication, otherwise the connection is terminated. CHAP is considered much more robust because the ID and Password do not travel across the link in clear text, and because the authentication may happen multiple times after the link is established.

IP Control Protocol (IPCP), documented in RFC 1332, is a PPP Network Control Protocol (NCP) used for establishing and configuring IP over PPP. IPCP negotiates the IP addresses to be used, and also negotiates other options such as the use of TCP/IP header compression with PPP. IPCP negotiation usually occurs after any authentication has been successfully completed.

Primary Network Concepts

The invention of the telegraph spurred the deployment of the first national, international and intercontinental networks of telecommunication cables. Several decades later, this same type of network was adopted for telephony that was invented by Alexander Graham Bell. He replaced binary code of Morse with a signal that is much more familiar to most of us, and that was *Voice*.

At first, making a phone call from one building to another required that a pair of wires be run from one to the other. To add a third building to this primitive network, required that it be linked to both of the first two. This meshed network topology quickly became inadequate and telephone systems soon adopted one of the fundamental networking concepts: *switching*.

Switching is a simple concept that consists of establishing or modifying connections between any two points of a network. Its basis is the need to take advantage of a common infrastructure under the assumption that it is highly unlikely that all users will be connected simultaneously. Thus telephony gave rise to another key concept in modern networking – *Sharing*.

The telephone system is based on *circuit switching*, a technique allowing a dedicated electronic circuit linking two remote terminals to be established or terminated. When the switched circuit is open, it can be used for transmitting voice by telephone, photocopies by facsimile, or any data in digital form by modem. The devices that manage the connections required for opening or closing circuits are called *Switches*. A company that operates such a network has an important advantage over the end user. It designs, controls and charges for the basic products of the network, i.e. its services. For the user, these services are usually indistinguishable from the network itself.

At every end of a telephone line, a terminal i.e. the telephone, allows the user to initiate or receive communications. The relationship between the central system and this terminal is of the master-slave type. So, the essential intelligence of the telephone network resides with the network itself, not with the terminal, which is only an instrument in the connection process. This whole architecture is designed and deployed in order to generate the end product of telecommunications companies – *Revenues*.

Another type of network appeared in the 20th century leading to the emergence of mass media. These are broadcasting networks, most notably television and radio, and since these network users are essentially passive, we will not discuss them here.

The dictionary defines a network as a "group of computers and associated devices that are connected by communications facilities." Thus, a network can be anything from two computers connected by a serial cable to thousands of computers connected by high-speed data communication links dispersed throughout the world.

A network provides two principle benefits:

1. The ability to communicate.

2. The ability to share.

A network supports communication among users in ways that other media cannot. E-mail, the most popular form of network communication, provides low-cost, printable correspondence with the capability for forwarding, acknowledgment, storage, retrieval, and attachments. Sharing involves not only information (database records, e-mail, graphics, etc.), but also resources such as; applications, printers, modems, disk space, scanners, etc. Through its ability to share, a network promotes collaboration. This is the main attraction of popular software called Groupware that is designed to allow multiple users to hold electronic meetings and work concurrently on projects.

In the midst of the rapid burgeoning of digital technologies in the last half of the 20th century, other kinds of network switching appeared. The main ones are frame relay, cell switching (ATM or SONET technologies), and *Packet switching* – the most famous of all and at the very foundation of the Internet and IP technologies in general.

In an IP network, messages are broken down into small packets of data, also known as Data-grams, which is analogous with the telegram. Each packet consists of a small part of the message plus a block of technical information including the addresses of origin and destination and the order in which to reassemble the packets. This block of information is called the *Header.*

Generally speaking, packets are transmitted from one point to another using the shortest path between the points at any given moment. At each intersection a piece of equipment is responsible for analyzing the header and deciding, in a fraction of a millisecond, which route the packet will take to arrive at its destination. These devices are called *Routers*.

The route taken by Packet-A can therefore be different from that taken by packet B. Moreover, the packets can arrive in any order; it is the system on the receiving end that is responsible for sorting them into the proper order. This is the defining characteristic of transmissions in connectionless mode, also known as *Data-gram Mode*.

For one reason or another, if a packet gets lost in the route, the receiving machine is alerted to the fact by the headers of the preceding and following packets. It can then send a message to the originating machine, asking for the missing packet to be re-sent. Such messages are repeated until the missing packet arrives. This process is called *Error Correction.*

The architectural differences between circuit switching and packet switching networks are substantial. In the first type, the user's message is channeled directly through a circuit established by an intelligent network. The owner of that network, in turn, specifies the nature of the services offered and bills for them accordingly. In general, this is called a *Centralized Network*.

In the case of IP networks, messages follow random paths and the intelligence required to determine their nature and ensure their integrity resides at the terminal points of the network, i.e. with the users' machines. The network's one simple task is to complete the physical transfer of data. It is not guaranteed that every data packet will reach its destination, and it is not important that it do so as long as other means are deployed in order to obtain the desired results.

This basic feature underlies what we call *End-to-end IP networks*. One of the primary benefits of this principle is that end-to-end IP networks are more flexible and more scalable and much more economical than circuit-switching networks.

What is a LAN (Local Area Network)?

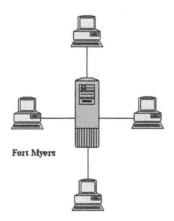

Fort Myers

A LAN is a single network of computers, which are physically located in the same area. An example of a LAN would be a network found in a Small Office/Home Office (SOHO) scenario. A LAN is usually comprised of small numbers of computers and printers with up to five servers. Local Area Networks seldom span over more than one location. The words Local Area are used in LAN because these networks of computers are usually in one building, even in one room, as opposed to computers that share information from one city to another. Local Area Networks (LAN's) allow you to do the two main things you do with computers, only easier and faster: 1) Share information and 2) Print documents. Most LAN's are run on what is called the Ethernet protocol. Basically, a LAN is a type of network, and Ethernet is a type of LAN – the most common type. IBM's Token-Ring and Apple's AppleTalk are other types.

Components of a LAN – Hardware and Software

❖ **Hardware:** There are two main types of LAN hardware: Cabled and Wireless. The choice between these two systems can decide how you physically set up your LAN. The current standard is what is called RJ-45 cable (CAT5, CAT5e and CAT6), which can support 10-Base T and 100-base Tnetworking. The 10 and 100 stand for the Megahertz or millions of cycles per second throughput of the LAN. This 8-pin wiring is similar to the standard four-pin phone cable and can actually be used for phone connections as well. Both telephone and ethernet connectors are easily assembled using a crimper tool, which can be bought at most discount stores. Networks can be costly because they require additional hardware devices called hubs that handle the traffic from one machine to the other. However, a modern hub can be purchased for as little as $30. Whether you buy wireless or RJ-45 cable, you will need a Network Interface Card (NIC) that is installed in your computer. In the case of cabled networks, you attach this cable to the NIC. A lot of the newer computers already have network cards in them, so you should double check before you make your final purchase. Also, you might need a specific card for a specific machine, so you may want to check with the vendor before you buy a network card for your computer.

❖ **Software:** This the most complicated part of a LAN, and the one that requires at least some minimal training. Macintosh systems and Windows 95/98 have built in LAN (networking) software, but there are many other name brands of LAN software – including Windows NT, Novell, and LANtastic.

The software usually allows the LAN administrator to make different folders accessible to different users, so you might have an accounting folder and a development folder, as well as providing some printing controls, such as making users able to choose between different printers, and deferring or canceling print jobs, etc.

❖ **Layout:** The physical way in which you lay out your cables depends on the cables you buy. Coaxial cable (like your cable TV) runs from one computer to the next, to the next, to a printer perhaps, and keeps going. It is a single line connecting all the computers (i.e. called a Daisy-chain). This is a good set-up if you have a few computers that are not going anywhere, or you are planning on running these cables over a long distance of over 400 feet). The main problem with this set up is that if you want to add a computer or printer to the LAN, you have to either cut or unplug the cable, add a new cable going to the new machine, and add a new cable going back to the previous line. This will disrupt the LAN, which means it should be done after hours; and you usually ends up using a lot more cable than you had originally planned. The RJ-45 cable (like the phone cable) is set up in what is called a Star model because it has a single cable going from each computer and each printer to a central hub, like an octopus with a machine on the end of each of its tentacles. However, this means that you need to purchase a hub, or even more than one, to set up your LAN. Hubs act as a Grand Central Station where information coming from one computer gets rerouted through the hub to another computer. Hubs generally connect anywhere from 4 to 18 computers at a time. To connect more machines, or if you have multiple floors, you can run a cable that connects two hubs together. So now you have two octopi connected by one of their limbs. Of course, you do not need to run the cable straight from the computer to the hub – the cables can go under desks, around corners, or to an out-of-the way closet where little hubs sit with their blinking lights, thinking deep thoughts. The benefit with this set-up is that it is easier to change. You can unplug one computer, and the rest stay connected to the hub, so the LAN still works.

❖ **Print Sharing:** One of a network's basic purposes is for several computers to share printers. This might require you to specify a particular computer, which will act as the traffic cop for files that need to be printed. Generally, once you have set up the print sharing, you do not need to regularly modify it, unless you are adding new printers or computers. Most LAN software comes with print sharing tools that will allow you to cancel print jobs, or delay print jobs for a later time; especially important if you accidentally printed a 100 page volunteer manual right before someone wanted to print their grant proposal on a deadline.

❖ **File Sharing:** Another basic function of a LAN is to allow computer users to share or copy files from one computer to another, without having to put it onto a diskette. Usually, to get into another computer over the network, regardless of whether you are on a peer-to-peer system or a server-based system, you need to log-on to that computer, i.e. you need an ID and a password which gives you access to that computer. If the files are scattered in different computers, you can imagine how quickly things would get complicated if every computer had an ID and password for every other person. Therefore, if you are sharing a lot of files, or if several people regularly work on documents together, it is always a good idea to have a central folder where these files are commonly kept. In small LAN's, especially peer-to-peer ones, it is easy to set up a folder on someone's computer - called the shared folder. On larger LAN's, you would set up the server computer to do the file and print sharing. In any case, you now only need to set up ID's and passwords on that central computer. You can go one step further and give certain people privileges to special folders, and restrict other people's access to that folder. The typical example is sensitive financial documents that only accountants and managers would be given access to see. Be careful though, because as a LAN administrator, it is your job to keep track of people's ID's, passwords, who has access to what, and the more restrictions you have in place, the more difficult it is to keep track.

What is WAN (Wide Area Network)?

A WAN is a network of LAN's. WAN's span more than one geographical area and are used to connect remote offices to each other. Basically, a WAN comprises of two or more LAN's joined together by routers. Routers are hardware devices that direct traffic from one LAN to another.

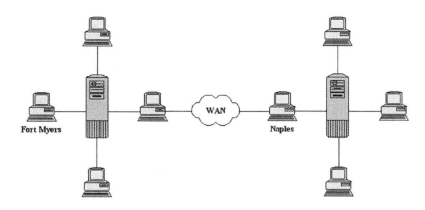

While wireless LAN's are used to allow network users to be mobile within a small fixed area, wireless WAN's are used to give Internet connectivity over a much broader coverage area, for mobile users such as business travelers or field service technicians. Wireless WAN's allows users to have access to the Internet, e-mail, and corporate applications and information even while away from their office. A WAN would be found in medium to large sized businesses with more than one office location. For example, a software company might have its headquarters in Dallas, but also have remote office locations in Austin and Ft. Worth. The LAN in the Dallas office would be connected to the LAN in the remote offices forming a WAN.

Wireless WAN's use cellular networks for data transmission and examples of the cellular systems that are used are: CDMA, GSM,

GPRS, and CDPD. A portable computer with a wireless WAN modem connects to a base station on the wireless networks via radio waves. The radio tower then carries the signal to a mobile switching center, where the data is passed on to the appropriate network. Using the wireless service provider's connection to the Internet, data communications are established to an organization's existing network. Wireless WAN's use existing cellular telephone networks, so there is also the option of making voice calls over a wireless WAN. Both cellular telephones and wireless WAN PC Cards have the ability to make voice calls, as well as pass data traffic on wireless WAN networks.

LAN's and WAN's come in many different flavors. The most popular type of network is Ethernet. Ethernet networks have speeds of 10 Mbps, 100 Mbps, or 1 Gbps. A 10 Mbps Ethernet network transmits data at 10 million bits per second. A 100 Mbps Ethernet network transmits data at 100 million bits per second. A 1 Gbps Ethernet network transmits data at 1000 million bits per second. The majority of networks today operate at 100 Mbps. 1 Gbps networks, however, are becoming more common as technology and bandwidth demand increases. An alternative to Ethernet networks is Token Ring networks. Token Ring networks operate at 4 Mbps or 16 Mbps and are usually found in legacy systems / networks (older systems / networks that have not been upgraded to newer technology). Token Ring networks are seldom found in companies today because of their lack of speed.

LAN's vs. WAN's

The official definition of a WAN is a "communications network that connects geographically separate areas." However, the rub is trying to pin down what constitutes geographically separate areas.

In general the line between LAN and WAN is crossed once you start using telecommunications systems to connect the various segments of your network. Telecommunication systems include equipment and infrastructure provided and maintained by a third party, typically the phone company. For example, if your network uses communication devices to connect separate segments via public phone, ISDN or T1 lines, then you can probably call it a WAN. Still, this definition is not perfect. A single employee dialing into a RAS server from his laptop at home doesn't really constitute a WAN. However, there are few situations where the distinction between a LAN and WAN is so critical that you need to make a precise call in either one direction or the other.

Comparison between LAN's & WAN's

- **Coverage:** Wireless local area networks by definition operate over a small local coverage area, normally about 100 miles in range. They are typically used in buildings to replace an existing wired Ethernet, or in a home to allow multiple users access to the same Internet connection. Other wireless LAN coverage areas can include public hotspots in coffee shops or some city neighborhoods. Wireless wide area networks cover a much larger area, such as wherever the cellular network provider has wireless coverage. Typically this is on a regional, nationwide, or even global scale. Using a wireless WAN usually gives the user access to data wherever they go and is one of the biggest advantages of a wide area network.

- **Speed:** The 802.11b wireless LAN standard transfers data at speeds of up to 11 Mbps, with typical rates of between 1–4 Mbps, decreasing as more users share the same wireless LAN connection. The next version, 802.11a, is supposed to transfer data at speeds of up to 54 Mbps. However, a potential problem for throughput is overcrowding of the bandwidth. Many people or businesses using wireless LAN's in the same area can overcrowd the frequency band on which they are transmitting. Problems with signal interference are already occurring and airwaves may become over-crowded. Wireless WAN speeds differ depending on the technology used. GPRS networks offer a maximum user data rate of over 115 Kbps if all eight timeslots in a cell are allocated for data transmission. One timeslot can provide between 9 and 21 Kbps. However, a realistic and consistent user data throughput rate of 30–50

Kbps is expected and seen in practice when 4 timeslots are used, as currently supported by most networks. This may be increasing in the future. These timeslots are shared with the voice traffic on the GPRS network. Data speeds on CDMA networks were initially available at speeds of 14.4 Kbps, but have increased to a maximum throughput of 153 Kbps as carriers have implemented CDMA2000 1X (1xRTT) networks. This gives the user typical throughput speeds of 40–70 Kbps, in addition to doubling the voice capacity of the carriers network. Future wireless WAN technologies, like CDMA2000 1xEV-DO, provide peak data rates of up to 2.4 Mbps in a standard 1.25 MHz CDMA channel. UMTS, also known as WCDMA (Wideband CDMA) is another approved next generation standard, which utilizes one 5 MHz channel for both voice and data, offering data speeds up to 2 Mbps.

- **Data Security:** Security is one of the most important features when using a wireless network. Security is one of the biggest strengths for cellular wireless networks (WWAN's) and one of the biggest weaknesses in 802.11 networks (WLAN's). 802.11b networks have several layers of security; however there are weaknesses in all of these security features. The first level of security is to have wireless LAN authentication done using the wireless adapter's hardware (MAC) address. However, this alone is not secure because the MAC address of a wireless client can easily be falsely created. Security can be increased on wireless LAN's by using shared key authentication. This shared key must be delivered through a secure method other than the '802.11' connection. In practice, this key is manually configured on the access point and client, which is not efficient on a large network with many users. This shared key authentication is not considered secure and is not recommended to ensure security. Another weakness in an 802.11 network is the difficulty in restricting physical access to the network, because anyone within range of a wireless access point can send, receive, or intercept frames. Wired Equivalency Protocol (WEP) was designed to provide security equivalent to a wired network by encrypting the data sent between a wireless client and an access point.

- **Costs:** Since wireless LAN's operate in the unlicensed frequency range, there is no service cost for using a private wireless LAN, such as in a corporate office or home office. There will be a monthly Internet service provider cost for accessing the Internet through your wireless LAN access point through your broadband or cable connection. The other main cost involved is the cost of purchasing and installing the wireless LAN equipment and devices, and the cost of maintaining the network and the users. There are normally fees for using public 'hotspot' access. For cellular wireless WAN's, the wireless network is acting as your Internet service provider by providing access to the Internet over their wireless network. The wireless provider therefore charges a monthly subscription rate to their network, similar to a wireless phone subscription. This may be a flat monthly fee, for time connected to the network, or per megabyte of data transferred.

	Wireless LAN	Wireless WAN
Coverage	Office Buildings or Campus with some Public hotspots	Available wherever there is cellular network coverage; nationwide and global
Throughput Speeds	1-5 Mbps (However the underlying Internet Connection may yield a slower speed)	30-50 Kbps (GPRS) 40-70 Kbps (CDMA2000 1X)
Security	Security flaws	Secure encryption and authentication
Airtime Charges	Airtime charges exist for most Hotspot access. No airtime charges for office or home users (although ISP monthly service fee still exists)	Monthly subscription from wireless network provider
Uses	❖ Accessing a shared network within a building or across a campus	❖ Remote access to a corporate network for e-mail and applications ❖ Web and Internet access.
Voice	No	Yes
Wired Analogy	Ethernet Network	Remote modem access

Advantages	❖ High Speed ❖ No airtime charges to set up networks (hardware costs and broadband Internet connection fees still apply)	❖ Ubiquitous coverage ❖ Secure Network ❖ Access your data from anywhere
Disadvantages	❖ Localized coverage only ❖ Security problems	❖ Data rates faster than dial up, but not at wireless LAN speeds yet.

When considering wireless LAN and wireless WAN technologies, it is important to note the differences between them and ensure that you choose the right technology for your specific application. Both of these wireless technologies have great advantages when used in the right application, and can compliment each other when used together.

Can wireless WAN's and wireless LAN's work together?

Although wireless LAN's and wireless WAN's may appear to be competing technologies, they are far more useful as complementary technologies. Used together, a user would have the best of both technologies, offering high-speed wireless access in a campus area, and access to all their data and applications with high-speed cellular access from anywhere with wireless WAN network coverage.

A wired analogy of these complimentary technologies would be as follows:

A user would plug their laptop (with in-built network adapter) into a wired LAN connection while they are in the office. This gives them high-speed access to their e-mail, applications, data, and the web.

When they leave the office and work from home, or on the road at their hotel, they would use their dial up modem to have remote access to their e-mail, applications, and the web.

In the wireless example, the same user has a laptop with built in wireless LAN access. This wireless LAN access is used for high-speed access to applications while in the office. Once out of the office traveling to a local customer site, completing a work order in the field, or accessing e-mail from a hotel or airport, there is no longer any access to an 802.11 network. The wireless WAN card is now used to access a cellular provider's network and obtain secure, remote access to e-mail, applications, and the web.

Since many computers are now coming with wireless LAN devices built in, having a wireless WAN PC Card inserted into the computer would ensure that users can have high-speed wireless access where it is available, but still be able to access their important data with their wireless WAN card wherever there is cellular network coverage.

Transmission Channels and Switching Systems

The Internet backbone is the national and international network of large access points that link the Internet worldwide. The backbone consists of large switches linked together by high-bandwidth, and high-speed digital lines. These large switching points, called Network Access Points (NAP's) are a collection point and a Grand Central Station for Internet Service providers (ISP's) to access the Internet. They provide Internet access to ISP's. It is the ISP's that further distribute Internet access to end-users. ISP's connect to the NAP's using primarily OC12, OC3 ATM and DS3 ATM.

Internet servers

A single physical server may provide one or more of the services listed below:

HTTP (Web) Servers

Web servers are designed to use HTTP to move hypertext files across the Internet. The engine of the Web server is called an HTTP daemon. The daemon waits for HTTP requests from clients and handles requests when they arrive. The establishment of an HTTP connection by the client to a server usually uses the well-known port 80, the default port used for HTTP. HTTP 1.1 supports persistent connection. This means that once a browser connects to a Web server, it can receive multiple files through the same connection. This improves performance by up to 20%. Browsers supporting HTTP 1.1 will compress HTML files for transport across the Internet. This also provides a substantial savings in the amount of data that must be transmitted. HTTP 1.1 provides the ability to have multiple domain names share the same Internet address. This allows Web servers to host a number of Web sites. A Web site that does not have its own IP address is called a Virtual Server.

FTP Servers

FTP servers listen on the well-known port number 21 for control connection requests. FTP servers store and retrieve files that usually are organized by subjects. FTP servers are designed for quick, reliable file uploads and downloads between the server and an FTP client. You can FTP from the DOS command line or through a FTP-program. Vital commands are open files (to establish contact), get files (to download files) and put files (to upload).

Mail Servers

The primary function of a mail server is to receive and transmit e-mail from multiple users using e-mail clients. Popular e-mail servers include Microsoft's Exchange Server, Novell's Group Wise, ccMail and UNIX's sendmail.

Telnet Servers

Telnet is a terminal emulation program. Once you are logged on to a network, you can use a terminal emulation program, e.g. 'PC

Anywhere', to control a remote server as though you were a user logged on locally.

List Servers

List servers are used to maintain subscription e-mail mailing lists and to distribute information to the e-mail addresses on the mailing list, e.g. newsletters (Push-method). Lists servers content may also come from subscribers. Many list-servers allow subscribers to send e-mail to a specified e-mail address for distribution to the whole group.

News Servers

News servers support the distribution of news from users. Subject indexes, and a string of posted articles (posts) collectively are called a newsgroup. News servers distribute news content to subscribers. It is a function similar to that of list servers except that news servers operate on the pull model. No content is distributed unless the subscriber connects to the news server. The content is downloaded to the user's Network News Transport Protocol (NNTP) viewer. The NNTP reader accesses the news server at port 119.

LDAP (Directory) Servers

Directory servers hold indexed databases of categorized information that users can query. Two popular uses of directory servers are indexing Web sites by subject and indexing e-mail addresses and other personal information into white pages. On the Web, some directory servers index Web sites by keywords; others like Yahoo use a hierarchical tree of topics and subtopics. The other major use of a directory server is to develop indexes of directory information, such as e-mail addresses and public keys using Lightweight Directory Access Protocol (LDAP). LDAP is a simplified version of the X.500 protocol.

Mirrored Servers

Mirrored servers are servers that maintain copies of the same files as another server and are typically used to backup primary servers.

If the primary server fails, the mirrored server continues to operate without any downtime. They can also be used to spread the load to more than one site.

Certificate Servers

Certificate servers are designed to issue certificates (ID's). You must request a certificate for your organization through a Certificate Authority (CA) Secure Sockets Layer (SSL) uses a public and private key encryption system that also incorporates the use of digital certificates. The most widely used standard for defining digital certificates is X.509.

Proxy Servers

Proxy servers are designed to sit between a client computer, such as a browser, and a Web server. A proxy server intercepts all requests to the Web server to see if it can fulfill the request itself, but otherwise forwards the request to the Web server. The proxy server caches Web pages from an Internet Web server and acts as an intermediary or buffer between a Web client and a Web server. Proxy servers are also gateways that allow direct Internet access from behind a firewall. Proxy servers open a socket on the server and allow communication to the Internet via that socket. Proxy servers filter requests, provides security, and improves performance. An IP proxy masks the IP address of internal hosts and represents itself instead.

Mail Systems

Simple Mail Transfer Protocol (SMTP) is one of the most widely used protocols on the Internet. It was designed to handle the transfer of messages from one host to another. SMTP is not a mail system, only mail transfer protocol. It provides a mechanism for transferring messages between hosts, for queuing messages until a message can be forwarded (store and forward) and for notifying the sender when the mail cannot be delivered. By default SMTP uses port 25.

Web Systems

o **TCP/IP** is the protocol you use when connecting your network to the Internet. DNS is a client/server protocol that translates TCP/IP host names into numeric IP addresses. DNS maintains a set of tables that map host names to IP addresses.

o **'WINS'** is an automated NetBIOS name resolution method. WINS clients automatically register themselves with the WINS server, which can be queried for name resolution. NetBIOS names can also be manually entered into the database.

o **Network News Transport Protocol (NNTP)** services allow your server to act as a news server by providing newsgroup services to NNTP clients. Default port is 119.

o **IP Filtering** is granting or denying access to specific IP addresses by specifying a single client IP address, a group of addresses using a subnet mask, or a domain name.

Internet Addressing

o **Packet:** Information being sent across the network is referred to as packets. A message will often contain several packets that must be delivered in order with the data intact.

o **Protocol:** A set of rules defining how two processes communicate.

o **Host:** It is also referred to as an end node. A device configured as part of the network. The destination for an IP packet.

o **Address:** Each node must have a unique address to communicate via TCP/IP. The address must follow the IP address format.

o **Names:** Each host will also have a unique name. A common method for managing names is DNS. DNS servers map up IP addresses to DNS names.

Routers

Routers connect networks together using the network portion of the IP address to identify the appropriate sub network. A router reads a packet from one network, determines whether that packet is destined for another network, and forwards the packet appropriately.

IP Addressing

o   IP4, the current version of IP, uses a 32-bit address scheme that allows for 4.294.967.296 or 232 IP addresses.

o   IP addresses are 32 bits long

o   IP addresses are divided into four octets.

o   Each octet contains 8 bits with values ranging from 0-256

o   A zero or one represents a bit.

Position	8	7	6	5	4	3	2	1
Power of 2	2^7	2^6	2^5	2^4	2^3	2^2	2^1	2^0
Decimal value	128	64	32	16	8	4	2	1
Example	1	1	0	0	0	0	0	1
Equals	128	64	0	0	0	0	0	1
Total	128 + 64 + 1 = 193							

So an IP address that numerically reads: 196.250.28.3 will binary read:

11000100.11111010.00011100.00000011

Internet Hardware

o   Network Adapter Cards: Also referred to as a Network Interface Card (NIC), provides a communication channel between your computer's motherboard and the network. The Media Access Control (MAC) addresses are your NIC's unique hardware identifier, and are hexadecimal numbers that are assigned to each NIC card during the manufacturing process.

Bridges

o   Provide a way of segmenting network traffic and connecting different types of networks (different LAN types).

o   Operates at the Data Link Layer, or more precisely, on the MAC sub layer.

o   Can filter traffic based on addresses.

o   Cleans and amplifies signals allowing for expanded networks.

o   Modern bridges are usually referred to as learning (transparent) bridges because they are capable of automatically identifying devices on the segments they connect.

o   Bridge filtering consists of looking for patterns within the frame to selectively control frames that will be forwarded.

o   Layer 2 Switch (data switch) is generally a more modern term for multi-port bridge.

Routers

o   Allow you to build an inter-network computing environment and are a key element in wide area networking.

o   Operate on the Network Layer.

o   When a router receives a packet, it will generally forward it to the appropriate network, based on a table maintained in the router. The tables may be static or dynamic.

o   The router provides a port of entry that can control the entrance and exit traffic to and from a subnet.

o   Routers should be given preference over bridges when designing and configuring WAN's.

Brouters

Brouters operate on both the Network Layer for routable protocols and at the Data Link Layer for non-routable protocols. A combined bridge and router.

Firewalls

A firewall is a mechanism for controlling access between networks. Provides an extra layer of security to protect private organizational systems from external intruders and can be hardware based or software based. Typically, firewalls are implemented within the router that connects the private network with the public network. It is also possible for firewalls to filter and deny access to Websites that are considered inappropriate.

Packet Filtering

It is the technique of examining each data-gram as it passes through a router. If the contents of the data-gram agree with criteria defined by the security administrator and stored on the router, then the data-gram is passed on to its destination.

Bastion Hosts

Bastion Hosts are heavily fortified servers on the network that all external traffic must pass through. All services except those absolutely essential to running the system are eliminated. In this way, even if an intruder were to break into the system, damage would be limited.

Remote Access and Diagnostics

o   Ping: The Packet Internet Groper is used to test TCP/IP connectivity by transmitting Internet Control Message Protocol (ICMP) packets to a TCP/IP host. The host should then echo the packets back to the original IP address. Test for presence of other systems and ping yourself with the loopback address 127.0.0.1 to determine if you have a working TCP/IP stack.

o   IPConfig: The Internet Protocol Configuration utility is used to view TCP/IP configuration information from a command prompt, including IP address, subnet mask, and default gateway.

o   Tracert: Is used to check the availability of routes to a given destination network. It also provides timing information that can be used to identify bottlenecks in an Internetwork or on the Internet. Determines the route packets take to reach the specified destination.

ARP

o   Address Resolution Protocol and Reverse Address Resolution Protocol (ARP/RARP) are maintenance protocols. They are used on LAN's to enable hosts to translate IP addressing to the low-level MAC addresses.

o   ARP is used to request a station's MAC address when only the IP address is known.

o   RARP is used when the MAC address is known but not the IP address.

Remote Access Protocols

o   Serial Line Internet Protocol (SLIP): SLIP is a communications protocol that is part of the TCP/IP suite of programs. The protocol was devised to handle communications over fairly slow links (< 19.2 Kbps). SLIP is employed over telephone lines and requires minimal overhead. Disadvantages include inability to provide packet addressing, lack of error correction, lack of data compression, no flow control, no security, and not supporting IP addressing.

o Point-to-Point Protocol (PPP): PPP provides a dial-up connection. It can be used with protocols other than TCP/IP. It offers error correction, supports dynamic IP addressing, and password logons.

o Multi-link PPP: It enables two or more modems or ISDN devices to be combined as a single dial-up link that provides a bandwidth equal to the combination of modems. Multi-link PPP increases modem speed when using more than one modem/ISDN.

o Point-to-Point Tunneling Protocol (PPTP): It supports all PPP features and allows for secure connections over the Internet by tunneling other protocols within TCP/IP packets. PPTP data is encrypted and encapsulated inside PPP packets. L2F from Cisco is the alternative to PPTP. It is used to create VPN.

o Point-to-Multipoint Protocols: Point-to-multipoint protocols broadcast data from one sender to multiple points rather than just transmitting data to the intended recipient. On a point-to-multipoint network, a recipient's connection device is set to monitor certain bandwidth or look for certain addressing information in data packets. The connection device ignores the other bandwidths and data packets.

Internet Bandwidth Linking Technologies

o Synchronous Optical Network and Synchronous Digital Hierarchy (SONET): This is used for Internet backbones and for connecting large, public WAN communication systems.

STM 64 = 10 Gbps - backbone

STM-16 = 2,488 Gbps - Internet backbone

STM-4 = 622, 08 Mbps - Internet backbone

STM-1 = 155, 52 Mbps - Large company backbone and Internet backbone.

o   Switched Multi-megabit Data Service (SMDS): It is Packet-switched technology that is used to connect LAN's, create MAN's and WAN's. It operation path range is up to 155 Mbps (OC3-3).

o   Optical Fiber: Fiber-optic cable is comprised of light-conducting glass encased in plastic fibers surrounded by a protective cladding and a durable outer sheath.

OC256 = 13,271 Gbps - backbone

OC192 = 10 Gbps - backbone

OC48 = 2,488 Gbps - Internet backbone

OC24 = 1,244 Gbps - Internet backbone

OC12 = 622,08 Mbps - Internet backbone

OC3-3 = 155,52 Mbps - Large company backbone and Internet backbone

OC1 = 51,84 Mbps - ISP to Internet, smaller Internet links

o   Asynchronous Transfer Mode (ATM): Packet switching network service that can transmit data in excess of 600 Mbps. ATM is the backbone of major telecommunications companies. It also provides full support for voice, video, data and fax.

o   Fast Ethernet and Gigabit Ethernet: Ethernet is a technology that allows data to be transmitted between computers at 10 Mbps (10baseT Ethernet), 100 Mbps (Fast Ethernet) or 1 Gbps (Gigabit Ethernet). Ethernet media varieties include thick coaxial, thin coaxial, twisted pair and fiber optic.

o   Fiber Distributed Data Interface (FDDI): OSI-compliant standard used by large, wide range, fiber-optic LAN's in large companies and in some larger ISP's to support thousands of end users. Its operation path range is between 100 - 200 Mbps bandwidth.

Layers, Protocols and Interfaces

A network's architecture can be described in two ways: 1) peer to peer, and 2) client-server. A peer-to-peer network is a grouping of personal computers that all share information between each other. Peer-to-peer networks are usually comprised of less than ten computers. This type of network fulfills the needs of users that require very little computer services and security. File storage is scattered among the computers, and security is extremely low. Peer-to-peer networks are very susceptible to hackers and other malicious users because there is no solid security policy enforced. A more organized approach to networking is client-server architecture. A client-server network is a network comprised of several workstations and one or more servers. Clients (users) log in to the server and gain access to their files. In client-server networks an administrator can control the privileges of each user. Files are stored centrally simplifying data backup. Security policies are implemented that protect users' information. Client-server architecture has become the de-facto standard in small, medium, and large businesses because of the advantages it offers.

A network's architecture consists of a coherent hierarchical set of layers, protocols and interfaces. Designing a network of this type presents two major kinds of problems:

- First, issues with the physical medium: the nature and physical characteristics of the cables used, their interconnection, the physical topology of the network as a whole, etc.

- Second, software aspects, pertaining to the logical structure of the network: the order and hierarchy of the protocols used and the definition of interfaces between each software layer.

The organization of a network in software layers characterized by the stacking of one protocol on another, was established early on as the simplest, most rational and, most importantly, most scalable way to design a network. The basic theoretical principle is very simple and can be summarized with the following diagram:

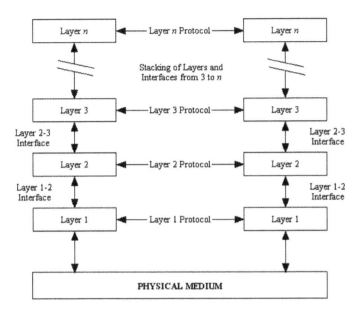

Even though numerical data travels physically from end to end in a single flow, in a logical sense, Layer 1 of the left host only communicates with Layer 1 of the right host. The "language" that allows them to communicate is called a protocol. A protocol consists of a set of specifications relating, for example, to the speed of communication, its coding, its starting and ending mechanisms, its error handling methods, etc.

Most of these layers are completely transparent to the user. Only network engineers usually deal with the lowest ones. Higher up the hierarchy, they are handled by the operating systems in order to allow different computer platforms to communicate with each other. Still higher up, the top layers are managed by software on the users' machines (browsing, chat, file transfer, multimedia, scientific, industrial, commercial and other applications).

Between each layer, a software interface makes it possible to define the basic operations and services that the next layer down makes available to the layer above. Just like protocols, interfaces play a critical role. If they are clear and well defined, they allow the replacement of one layer's protocol by another without having to modify the rest of the architecture. It is therefore protocols that allow the network architecture to evolve slowly, through updates or the implementation of new protocols.

Voice on Packets (VoIP)

IP telephony (IPT) sends voice conversations over the same network that carries data throughout your company, whether it is a local-area network (LAN), a wide-area network (WAN) or even the public Internet. IPT technology breaks the sound into tiny digital units called Packets[12]," then sends those packets over the network and reassembles them in the correct order on the receiving end.

From a technological perspective, sending voice on packets using Internet is broadly categorized under the term VoIP, i.e. Voice over Internet Protocol.

Voice over IP (VoIP), or Internet Telephony, is technology that uses data packets to transmit voice over the Internet. It consists of 3 steps. The conversion of the analog voice into IP packets is done by using a digital signal processor (DSP) embedded in the phone or using a PC in the case of soft phones, which are software programs used with a handset or headset connected to a PC.

IP packets over a packet or data network, where the packets travel along with other VoIP packets, data packets, Video over IP packets, etc., which are sent to appropriate destinations by routers and switches, require the conversion of IP packets back to voice (using DSP's or computers).

---

[12] The "packets" in this case are tiny pieces of voice and data that travel between sender and receiver in a network.

With the use of appropriate gateways (devices that do appropriate conversions), telephone calls can be made from a soft phone to a soft phone, soft phone to a POTS (plain old telephone system, i.e., regular) phone and vice versa, and POTS phone to a POTS.

The traditional voice network assigns a dedicated circuit to a telephone call for its duration. Although this produces very reliable telephone calls, assigning a dedicated circuit to each telephone call makes very inefficient use of bandwidth. VoIP uses compression techniques (to make very efficient use of available bandwidth. For example, by using the G.723 compression protocol, a VoIP call can be transported using one-tenth the bandwidth of a traditional voice call, which uses a 64 Kbs channel.

VocalTec, Inc first introduced Voice over Internet Protocol (VoIP) technology in February 1995. This innovative technology represents a new form of telephony in which voice and video transmissions can be digitized and transmitted over the Internet and other digital data networks. IP is very versatile, as it can travel over ATM, Ethernet, Frame Relay, ISDN and even analog lines. During the last four years, several enhancements have been made to improve the functionality, quality and reliability of VoIP technology.

There are multiple configurations for a VoIP implementation. For example, VoIP can be combined with an organization's existing infrastructure (PBX or PSTN) or function as a stand-alone unit. Most VoIP implementations currently involve the following components:

- Private Branch Exchange (PBX): The PBX is a switchboard system, which is responsible for routing intra-company calls and providing access to the external Public Switched Telephone Network (PSTN) for local and long-distance calls.

- Gateway: Voice signals to be transmitted over the packet-switched network (e.g., WAN) are routed from the PBX to the gateway, where the analog signal is processed, digitized and compressed into packets. The digital voice packets then travel to the network router.

- Router: The router acts as the connection between the company's internal network and another LAN or WAN (e.g., the Internet). Thus determining how the packets will be forwarded to their destination IP address.

Configurations – Local and Wide Area

Once the data leaves the router, it travels over the network to the remote site to which the original communication was directed. When the packets arrive at the router of the remote site, they are then forwarded to the gateway where they are reassembled for the final destination point. This process can be customized to work with phone-to-phone, phone-to-PC, and PC-to-PC communications. The following exhibit shows a common VoIP configuration and the Internetworking components between two sites.

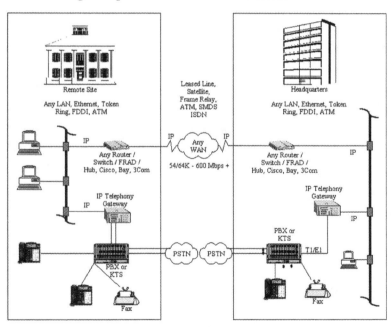

Source: MICOM Communications Corp. "Voice/Fax Over IP: Technology Overview": 13

It should be noted that current VoIP technology has not yet been perfected. For example, VoIP voice quality can vary greatly depending on network traffic and bandwidth availability. At its best, VoIP audio can be near toll quality; at its worst, it can result in unnatural or garbled speech. Lost packets and delivery delays usually cause these problems. Unlike data transmissions, voice transmissions are time-dependent and extremely sensitive to delay. In order for the speech to sound natural, these packets must be sent in a relatively constant stream with consistent arrival times. To deliver the best voice quality, a VoIP gateway must use a coder/decoder (codec) with good voice quality and low delay. Other features such as echo cancellation and forward error correction (FEC) can further reduce voice transmission difficulties.

Quality Issues – Voice Compression, Network Delay, Packet Loss, Echo & Jitter

VoIP packets must be delivered to the receiver in real time and with reliably. Sending VoIP packets over public IP networks, as opposed to private IP networks, typically generates voice connections of a very poor quality. This is because voice packets are intermingled with (typically low priority) data packets and both kinds of packets have to compete for bandwidth. Voice packets may be delayed that cause jitter. Network congestion may lead to some packets being dropped, which causes clicking and popping sounds. Likewise VoIP telephones where the computer is connected to the telephone and the telephone to the RJ45 utilizing just one CAT5 cable is subject to the same bandwidth conflict as above. Good VoIP implementations must alleviate all of these problems.

The public IP network was not designed for real-time packet transmission and typically does not reserve extra bandwidth to ensure that there are no transmission delays. Data traffic is subject to bursts and can be very large. In such cases, voice packets are delayed, as they have to compete for bandwidth with the data packets and may not be delivered in real time.

To address this issue, VoIP packets must be given a high priority as compared to data packets. In other words, voice packets should be given better quality of service (QoS). Several mechanisms for QoS have been proposed, such as:

- Diffserv (Differentiated Services) The plan redefines part of the existing Type-of-service (ToS) byte in every IP packet header to mark the priority or service level that packet requires.

- IPv6 (Internet Protocol Version 6) uses priority bits and flow labeling that tells the router on how to treat the packet, and RSVP (uses bandwidth reservation). The current version of IP is version 4, so it is sometimes referred to as Ipv4. IPng is designed as an evolutionary upgrade to the Internet Protocol and will, in fact, coexist with the older Ipv4 for sometime. IPng is designed to allow the Internet to grow steadily, both in terms of the number of hosts connected and the total amount of data traffic transmitted.

Sending voice and data over the same network facilitates the building of multimedia applications such as; converting voice mail to e-mail, unified messaging, and enables easier multi-media interactions with customers (simultaneous establishment of a voice and Web connection with a single click).

Use of VoIP telephony or simply IP telephony requires that the sender and receiver have one of the followings:

- IP phones (phones that convert voice to IP packets before sending them on to the network)

- Soft phones (Microsoft's Windows XP operating system will provide built-in support for IP telephony).

- Access to gateways that convert analog voice to IP packets and then back to voice – IP packets are transmitted between the gateways.

Advantages of VoIP

- Users benefit by new applications based on converged voice and data networks. For example, VoIP –

  o Eliminates long distance charges over LAN's and WAN's.

  o No need to maintain separate voice and data networks in the enterprise,

  o Enables web call centers,

  o Supports multi-media conferencing and collaboration, and

  o Provides unified messaging using a phone or PDA to access both e-mail and voice mail. Appropriate conversions are performed using text-to-speech and speech recognition systems.

- Service providers benefit with more efficient use of their networks. For example,

  o Voice packets can be sent along with data over a single network,

  o Compression techniques can be applied to send more voice packets over the same bandwidth thus enabling more VoIP connections (normal voice channels require 64 Kbs; this bandwidth can be reduced to 6.4 Kbps by using compression techniques such as G.723; further compression has a noticeable negative effect on voice quality), and

  o A single network can be used for transmitting both voice and data.

- Equipment makers will benefit from the new types of equipment and software. For example,

  o VoIP gateways for converting analog voice to IP packets and vice versa. This allows the use of the

PSTN for the last mile of connection to the phone users location.

    o   IP phones are the phones which convert voice to IP packets and vice versa, and an IP connection is needed for such phones. A voice connection is not required.

Uses of VoIP

VoIP will be used for example, in:

- Backhaul carrier networks,

    o   To avoid very high tariffs on normal international voice connections over the public telephone networks as many countries have very high telephone tariffs. VoIP packets go over the data networks and are not subject to voice telephony tariffs. Using VoIP over private networks can save up to 80% in costs.

- Converged data and voice networks,

- Converged data and voice appliances such as IP PBX's, and always-on cellular phones.

Some Standards Used in VoIP

G.711: It's an international standard used for encoding/packetizing telephone voice at either 56 Kbps or 64 Kbps. This is uncompressed digitized voice.

    o   G.723: It's a protocol for compressing voice to 6.4 Kbs or 5.3 Kbs. The compression quality is very good with voice quality, and as good as normal telephone voice quality. It is supported by virtually all IP telephone equipment.

    o   H.323: It's the signaling & telephone services protocol for the transmission of IP packets representing any combination of voice, video and data. H.323 is designed for operation over existing IP networks. It includes facilities call setup signaling and media control, and allows VoIP equipment to interoperate.

o   SIP (Session Initiation Protocol): This is a signaling & telephone services protocol similar to – but simpler than – H.323.

Although data and voice can share the same network, they do have different requirements. It is helpful to understand how voice differs from data in order to ensure that they 'play well' together on the same network.

Voice works in real time

Voice is described as a real-time application. Thus it's critical for the packets to flow smoothly. A few lost packets can noticeably degrade the quality of the communication. In order to ensure that a voice conversation maintains an acceptable quality level, voice must be guaranteed a certain amount of bandwidth on the network.

Voice doesn't necessarily have to preempt all the data traffic on the network, but it does need to be guaranteed a certain minimum amount of capacity. By incorporating a set of standards-based tools or mechanisms called Quality of Service, companies can make sure that both voice communications and mission-critical data applications perform well.

Voice is more predictable than data

Voice traffic is more orderly and predictable than data traffic. Data traffic often succumbs to 'bursts', which means subject to spikes of activity that can consume all of the available bandwidth, while voice traffic takes a predictable amount of bandwidth. Voice traffic is also symmetrical i.e., the packet size tends to be the same in both directions. Data traffic, on the other hand, is generally characterized by larger downloads and smaller uploads such as database queries.

Voice is less tolerant of delay

Further complications arise because voice is an application that cannot tolerate much of a delay. If there is too much traffic on the line, or if a voice packet gets stuck behind a large data packet, the voice packet will be delayed to the point that the quality of the call is compromised.

Latency is the average travel time it takes for a packet to reach its destination. The maximum amount of latency that a voice call can tolerate one way is 150 milliseconds (100 milliseconds is optimum).

Voice doesn't like variability

In order for voice to be intelligible, consecutive voice packets must arrive at regular intervals. 'Jitter' describes the degree of variability in packet arrivals, which can be caused by bursts of data traffic or just too much traffic on the line. Voice packets can tolerate only about 75 milliseconds (40 milliseconds is optimum) of jitter delay.

Voice doesn't like packet loss

Packet loss is a common occurrence in data networks, but computers and applications are designed to simply request a retransmission of lost packets. Dropped voice packets on the other hand, are discarded, not retransmitted. Voice traffic can tolerate less than a 3% loss of packets (1% is optimum) before callers begin experiencing disconcerting gaps in conversation.

Chapter Four
Basics Fundamentals of IP Telephony

---

a)   **Circuit Switching**

b)   **Packet Switching**

c)   **Protocols**

d)   **Dedicated, Switched Access and Virtual Circuits**

e)   **VPN's (Virtual Private Networks)**

f)   **Ethernet Networking**

Fundamentals of IP Telephony

The Internet has grown up to be the most influential factor guiding the convergence process at the present time, and the Internet protocol suite has become a shared standard that is used with almost any service. The Internet protocol suite is the combination of the Transport Control Protocol (TCP) and the Internet Protocol (IP); consequently, this commonly refers to the term, which encompasses the whole protocol family – TCP/IP.

In today's information society, IP-based networks have proved to be very important. This technology might appear a bit perplexing and crushing at the first look, therefore, we need to start by discussing the underlying network components based on which this technology is built.

A network has two basic and essential parts named links and nodes. Any type of network device like a computer is termed as a node, which are able to communicate with other nodes via links such as cables. Typically, the two different techniques for communication between nodes on a network are: the circuit-switched network and the packet-switched network. A traditional telephone system uses the later techniques, while IP-based networks use the former.

Circuit Switching

A closed circuit is created between two nodes in the network by a circuit-switched network to establish a connection. The established connection is thus dedicated to the communication between the two nodes. As almost no transmission uses the circuit for the entire time, wasted capacity is one of the immediate problems with dedicated circuits. More so, if a circuit fails during the transmission, the entire connection needs to be dropped and a new one has to be established.

As circuit-switching techniques are quite simple in carrying analog signals, it was the first used in communication networks. In this technique the transmission medium is usually divided into different channels with the use of Time Division Multiplexing (TDM), Frequency Division Multiplexing (FDM) or Code Division Multiplexing (CDM). A circuit is a string of concatenated channels from the source to the destination that facilitates information flow.

A signaling mechanism, which only carries control information is considered an overhead and is used to establish the circuits. Since the signaling process takes all decisions, it is also the most complex constituent in circuit switching. The general assumptions are that due to the signaling and per-circuit state management, circuit switches are hard to design, operate and configure.

The channel bandwidth is reserved for an information flow in circuit switching. The capacity of the circuit has to be at least equal to the peak transmission rate of the flow for ensuring timely delivery of the data. In this case the circuit is said to be peak allocated, and then the network offers a connection-oriented service with a perfect quality of service (QoS) in terms of bandwidth guarantees and delay jitter. However, in the process of achieving this, bandwidth is wasted when sources are idle or simply slow down.

Trouble occurs only at the time of allocating channels to circuits during circuit/call establishment. If there are not enough channels for the request, the call establishment may be delayed, blocked or even dropped. Contrary to it, once the call is accepted, resources are not shared with other flows, which ultimately results in eliminating any uncertainty, removing the need for buffering, processing or scheduling in the data path. When circuits are peak allocated, the only measure of Quality of Service (QoS) in circuit switching is the blocking probability of a call.

To sum up, circuit switching provides traffic engineering and traffic isolation, but at the cost of signaling overhead and using bandwidth inefficiently. Time and again it's said that circuit switching is made highly inflexible by these two drawbacks, specifically in an extremely dynamic environment like the Internet. I will argue in this that these drawbacks are outweighed by the advantages of using more circuit switching in the core of the network.

Packet Switching

On the other hand, networks based on Internet Protocol make use of a packet-switched network technology that efficiently uses the available capacity and reduces the risk of possible problems like disconnections. Messages sent over a packet-switched network are first divided into packets containing the destination address and after that each packet is transmitted through the network. In the process, every intermediate node and router in the network determines the next destination of the packet. A packet does not need to be routed over the same links as previous related packets. Packet transmission can remain successful even if packets travel over different routes, or in the event of a node malfunction or link breakdown.

In packet switching, which is the basis for the Internet Protocol (IP), information flows are disintegrated into various size packets, or fixed-size cells as in the case of ATM. These packets are sequentially sent to the nearest router that will search the destination address and forward them to the corresponding next hop. This process is repeated until the packet reaches its destination. The routing of the information is thus done locally, hop-by-hop. Routing decisions are independent of other decisions in the past and in other routers. However, they are based on network state and topology information that is exchanged among routers using IS-IS, BGP or OSPF. No other state is needed for the network operation except the routing tables.

The forwarding mechanism is called store-and-forward because while being processed, IP packets are received and stored entirely in the router and then transmitted. Additionally, packets may need to be buffered locally to resolve contention for resources. If the system runs out of buffers, packets are dropped.

With the most scheduling policies like WFQ and FCFS, packet switching remains work conserving. It keeps the link busy till the packets are available to be sent. Due to this, the statistical multiplexing gain is kept intact, i.e. the capacity of an outgoing link can be much smaller than the sum of its tributaries and still have a

117

packet delay or drop probability within certain statistical bounds. When the traffic is more rupturing the gain is higher. The main characteristics of packet switching are the statistical multiplexing and the buffering needs, as they are crucial in its comparison with circuit switching.

With the Internet, the network service is the best effort and connectionless, for it provides no delivery guarantees. Flow control, reliability and connection-oriented services are provided by end-to-end mechanisms like with TCP. As the underlying service is the best effort, there are no guarantees in terms of maximum delay, packet drops, bandwidth or delay jitter.

A great deal of research was done in the dawning period of computer networking, comparing circuit switching, message switching (a variant of packet switching, in which the whole information flow is treated as a single switching unit) and packet switching. Most of the research was done in the context of satellite, packet radio, and local area networks, which subsequently demonstrate how in these environments packet switching provided higher throughput on the average delay for a given bound. Packet switching not only made an effective use of the network bandwidth, but it also was robust and resilient to node and link failures.

Later research study on various scheduling algorithms and signaling mechanisms like Generalized Processor Sharing (GPS), Weighted Fair Queueing (WFQ), Integrated Services (IntServ), Differentiated Services (DiffServ) and Deficit Round Robin (DRR), explained how QoS guarantees can also be provided by packet switching on how the control of the admission of new flows to the network is achieved.

Protocols

As with many communications systems, the protocols involved in Voice over IP (VoIP) follow a layered hierarchy that can be compared with the theoretical model developed by the International Standards Organization. A system can be made more manageable and flexible by breaking a system into defined layers. Each layer has its own assignments, and doesn't require an elaborate perception of the surrounding layers. For example, Internet Protocol data-grams can be transported across a variety of link layer systems including Ethernet, serial lines (using PPP) and Token Ring. The link layer protocol is for the most part irrelevant to IP (unless that protocol limits the size of its data-grams), and need not be the same for the first link of a Voice over IP call and for the final link of a VoIP call.

There are always exceptions like IP over ATM, but the simple discreet layered model will be considered further in our discussion. The effect of the contribution of each layer in the communication process is an additional header preceding the information transmission. The complete packet, including header and data that a layer creates, becomes the data passed to the next level for processing. Those layers will then add a header portion, and it goes on like this. Each layer started at the Network/Internet Layer is considered for our discussion, and will be discussed in the sections lying ahead.

Internet Protocol (IP)

The lowest level protocol considered in our study is the Internet Protocol. IP is responsible for the transmission of packets/data-grams between host computers. Internet Protocol is called a connectionless protocol, as it doesn't establish a virtual connection through a network prior to commencing transmission, leaving this job for higher-level protocols. Internet Protocol (IP) gives no guarantees pertaining to reliability, error detection or error correction, and flow control resulting in data-grams possibly arriving at the destination computer out of sequence with errors, or even complete non-arrival. Nevertheless, Internet Protocol succeeds in making the network transparent to the upper layers involved in voice transmission through an IP based network.

Any Voice over Internet Protocol transmission must use IP by definition. IP is not well suited to voice transmission. Real time applications such as voice and video require guaranteed connection with consistent delay characteristics. These issues are addressed by higher layer protocols up to a certain extent.

The tabular illustration below shows a header, which advances the data payload to be transmitted. The header comprises 20 octets in its most basic form. There are optional fields that can be appended to the basic header, but these offer additional capabilities, which are not necessary for VoIP transmission.

	0 1 2 3 4 5 6 7 8 9 10 11 12 13 14 15 16 17 18 19 20 21 22 23 24 25 26 27 28 29 30 31			
	Octet 1, 5, 9, 11...	**Octet 2, 6, 10, 14...**	**Octet 3, 7, 11, 15...**	**Octet 4, 8, 12, 16...**
1 - 4	Version IHL	Type of service	Total length	
5 - 8	Identification		Flags	Fragment offset
9 - 12	Time to live	Protocol	Header checksum	
13 - 16	Source address			
17 - 20	Destination address			

The above-mentioned fields are briefly described below:

- **Version:** The version of IP being used; the version would be 4 for this format header.

- **IHL:** The length of the IP header is in the units of four octets (32 bits). For the basic header shown in the illustration, the value would be 5 with each line in the diagram representing four octets.

- **Type of service:** It specifies the quality of service requested by the host computer sending the data-gram. Routers or Internet Service Providers do not always effectively support this.

- **Total length:** This is the length of the data-gram, which is measured in octets; it includes the header and payload.

- **Identification:** Along with the handling of addressing data-grams between two host computers, Internet Protocol also needs to handle the splitting of data payloads into smaller packages. This process, which is also called fragmentation, is required because lower link layer protocols like Ethernet cannot always handle these large packet sizes, even though a single IP data-gram can handle a theoretical maximum length of 65,515 octets. This field is a unique reference number assigned by the sending host to aid in the reassembly of a fragmented data-gram.

- **Flags:** These flags indicate whether the data-gram is fragmented or not, and if it has been fragmented, whether further fragments will be following the current one.

- **Fragment offset:** This field indicates where this fragment exists in the data-gram. It is measured in units of 8 octets, i.e. 64 bits.

- **Time to live:** This field indicates the maximum time the data-gram is permitted to remain in the Internet system. This parameter ensures that a data-gram is given a finite lifetime, if it can't reach its destination host.

- **Protocol:** This indicates the higher-level protocol in use for this data-gram. Numbers are assigned for using this field to represent such transport layer protocols as TCP and UDP.

- **Header checksum:** This is a checksum covering only the header.

- **Source address:** The IP address of the host that generated this data-gram is also called source address. IPv4 addresses are 32 bits in length and when written or spoken, a dotted decimal notation (e.g.: 192.168.0.1) is used.

- **Destination address:** It is the IP address of the destination host.

User Data-gram Protocol (UDP)

By and large, there are two protocols available at the transport layer at the time the information is being transmitted through an IP network. They are User Data-gram Protocol (UDP) and Transmission Control Protocol (TCP). Both protocols enable the transmission of information between the correct processes, or applications on host computers. These processes are associated with unique port numbers, e.g. the HTTP application is usually associated with port

80. Transmission Control Protocol is a connection-oriented protocol, which establishes a communications path before transmitting data. It handles sequencing and error detection, and ensures that the destination application receives a reliable stream of data.

Voice is a real-time application, and mechanisms must be in place to ensure that information is received reliably and in the correct sequence, with predictable delay characteristics. Although Transmission Control Protocol (TCP) addresses these requirements up to a certain extent, there are some functions that are reserved for the layer above Transmission Control Protocol (TCP). Therefore, TCP is not used for the transport layer, and UDP – the alternative protocol – is commonly used.

Same as IP, UDP is also a connectionless protocol, which routes data to its correct destination port, but does not attempt to ensure data reliability or perform any sequencing.

	0 1 2 3 4 5 6 7 8 9 10 11 12 13 14 15 16 17 18 19 20 21 22 23 24 25 26 27 28 29 30 31			
	Octet 1,5	Octet 2,6	Octet 3,7	Octet 4,8
1 - 4	Source port		Destination port	
5 - 8	Length		Checksum	

The fields depicted in the table above are briefly described below:
- **Source port** identifies the higher layer process that originated the data.

- **Destination port** identifies with a higher layer process to which this data is being transmitted.

- The **length** in octets of the UDP data and payload – a minimum of 8.

- **Checksum** is an optional field supporting error detection

Real-time Transport Protocol (RTP)

In order to make sure that a stream of data is accurately reconstructed, real time applications need mechanisms to be in place, as data-grams must be reconstructed in the correct order. A means of detecting network delays is also required to be in place. Jitter is the variation in delay times experienced by the individual packets making up the data stream. In order to reduce the effects of jitter, data must be buffered at the receiving end of the link so that it can be played out at a constant rate. To support this requirement, two protocols have been developed. These are RTP Control Protocol (RTCP) and Real-time Transport Protocol (RTP).

RTCP provides feedback on the quality of the transmission link. RTP transports the digitized samples of real time information. RTP and RTCP do not reduce the overall delay of the real time information. Nor do they make any guarantees concerning quality of service. The RTP header that precedes the data payload is shown in the diagram below:

	0 1 2 3 4 5 6 7 8 9 10 11 12 13 14 15 16 17 18 19 20 21 22 23 24 25 26 27 28 29 30 31			
	Octet 1,5,9	**Octet 2,6,10**	**Octet 3,7,11**	**Octet 4,8,12**
1 - 4	V=2 P X CC	M PT	Sequence number	
5 - 8	Timestamp			
9 - 12	Synchronization source (SSRC) number			

The fields depicted in the table above are briefly described below:

- **Version:** Identifies the version of RTP (currently 2).

- **Padding:** A flag that indicates whether the packet has been appended with padding octets after the payload data.

- **X (or Header Extension):** Indicates whether an optional fixed length extension has been added to the RTP header.

- **CC (CSRC count):** Although not shown on this header diagram, the 12-octet header can be expanded optionally for including a list of contributing sources, which are added by mixers, and are only relevant for conferencing applications where elements of the data payload gets originated from different computers. CSRC's are not required for point-to-point communications.

- **Marker (M):** It allows significant events like frame boundaries to be marked in the packet stream.

- **PT (Payload type):** This field identifies the format of the RTP payload and determines its interpretation by the application.

- **Sequence number:** A unique reference number, which incrementally increases by one for each RTP packet sent. It allows the receiver to reconstruct the sender's packet sequence.

- **Timestamp:** The time that this packet was transmitted. This field allows the received to buffer and play-out the data in a continuous stream.

- **Synchronization source (SSRC) number:** A randomly chosen number that identifies the source of the data stream.

There are some other protocols that are listed below:

- **SGCP:** Simple Gateway Control Protocol. Controls Voice over IP gateways by an external call control element (labeled a call-agent). This has been adapted to allow SGCP to control switch ATM Circuit Emulation Service circuits (called endpoints in SGCP). The resulting system (call-agents and gateways) allows for the call-agent to engage in Common Channel Signaling (CCS) over a 64 Kbps CES circuit, governing the interconnection of bearer channels on the CES interface.

- **Media Gateway Control Protocol:** Media Gateway Control Protocol (MGCP) 1.0 is a protocol for the control of Voice over IP (VoIP) calls by external call-control elements known as media gateway controllers (MGC's) or call agents (CA's).

- **Skinny Client Control Protocol:** Skinny gateways are a series of digital gateways that include the DT-24+, the DT-30+, and the WS-X6608-x1 Catalyst voice module.

- **H.323:** Extension of ITU-T standard H.320 that enables videoconferencing over LAN's and other packet-switched networks, as well as video over the Internet.

Interconnecting VoIP Protocols

VoIP networks are popularly being deployed with an accelerated pace, with new & latest functionality being added frequently by VoIP service providers and vendors. As each protocol differs with supporting service by different vendors, and companies have varying business requirements, it is most likely for VoIP networks to continue to be made up of multiple protocols.

Customers get the required flexibility from various protocols to connect services from multiple carriers. Using standards, even multiple standards, still simplifies deployment of multi-vendor endpoints and increases options for network management and provisioning.

Companies are faced with several alternatives for interconnecting segments using differing VoIP protocols for the purpose of expanding their networks. These choices are segmented into any of these three categories:

- **Translation through Time Division Multiplexing (TDM):** In this model, a company uses either TDM equipment or VoIP gateways to translate from one protocol domain to another. The benefits of this model are that it can be used today. The downside is that it introduces latency into the VoIP network and involves yet another protocol translation. This model is generally considered as a short-term solution until IP based protocol translators are available.

- **Single Protocol Architecture:** In this model, a company takes all its VoIP devices and services to a single protocol and simplifies the network in totality. The disadvantage of this approach is that it might not be possible to migrate existing equipment to support the new protocol, which can limit the company's ability to take advantage of some existing services.

In addition, it limits the potential connectivity to other networks that are using other VoIP signaling protocols.

- **Protocol Translation:** Here with this model, companies use IP–based protocol translators to interconnect two or more VoIP protocol domains. A company can retain the flexibility of using multiple VoIP protocols with the support of IP translators. It doesn't introduce the delay problems that additional TDM interconnections do, and also a wholesale replacement or swap of existing equipment is not required.

The disadvantage of this approach is that there is no standard for protocol translation, so not all VoIP protocol translators exactly remain the same. Although the IETF has attempted to define a model for translating H.323 to SIP, it involves more than just building a protocol-translation box.

As shown in table below, protocols do have some differences even when they are somewhat similar.

	H.323	SIP	MGCP/H.248/ MEGACO
Standards Body	ITU	IETF	MGCP/ MEGACO— IETF; H.248— ITU
Architecture	Distributed	Distributed	Centralized
Current Version	H.323v4	RFC2543-bis07	MGCP 1.0, Megaco, H.248
Call control	Gatekeeper	Proxy/ Redirect Server	Call agent/ media gateway controller
Endpoints	Gateway, terminal	User agent	Media gateway
Signaling Transport	Transmission Control Protocol (TCP) or User Data-gram Protocol (UDP)	TCP or UDP	MGCP—UDP; Megaco/H.248—both
Multimedia capable	Yes	Yes	Yes
DTMF–relay transport	H.245 (signaling) or RFC 2833 (media)	RFC 2833 (media) or INFO (signaling)	Signaling or RFC 2833 (media)
Fax–relay transport	T.38	T.38	T.38
Supplemental services	Provided by endpoints or call control	Provided by endpoints or call control	Provided by call agent

Vendors of protocol translators need in-depth knowledge of all the protocols being used in the VoIP network, and they must be aware of how various VoIP components utilize different aspects of the protocol. As an example, SIP and H.323 can send Dual Tone Multi Frequency (DTMF) digits in either the signaling path or the media path (via RTP). However, H.323 mandates the use of only H.245 signaling path, and SIP doesn't specify how DTMF should be carried. This means that SIP devices could be sending DTMF in the media path (RFC 2833), and H.323 devices would send DTMF in the H.245 signaling path. If the VoIP protocol translator cannot properly recognize both the signaling path and the media path, then it might not function properly.

Based on the technical and business requirements at hand, companies will choose various protocols for their VoIP requirements, just as they choose various protocols for their data networks. Although the variety in VoIP protocols has caused some confusion in the marketplace, it is precisely this protocol flexibility that makes VoIP-based voice systems so much more useful than legacy voice systems. Three most important requirements that should be kept in mind while choosing vendors are:

1. Customers need products that support multiple protocols. This way, if a company finds that it needs to migrate its systems or add products that support a different protocol, it will not be required to perform upgrades to the network.

2. Customers need vendors that are committed to supporting open standards within their products and are actively developing voice strategies that consider interoperability with all VoIP protocols. Without this commitment, VoIP systems are in danger of becoming as proprietary as legacy voice systems.

3. Customers need voice solutions with end-to-end support for all VoIP protocols, meaning vendors must provide solutions that work in both single-protocol and multi-protocol environments.

Companies can aim at structuring dependable and scalable networks supporting the requirements of next-generation networks by working with vendors that can provide this VoIP flexibility.

Dedicated, Switched Access and Virtual Circuits

Dedicated Access

Dedicated Internet Protocol is a Quadrature Amplitude Modulation (QAM) and Quadrature Phase-shift Keying Transmission based in-band RF technology within the existing cable spectrum. Dedicated IP was visualized for allowing cable operators to deliver lucrative IP-based services like IP telephony to residential and commercial subscribers both. The dedicated Internet Protocol architecture and existing technologies both are present on the cable plant that also includes digital video, analog video and cable modem systems. The dedicated Internet Protocol architecture doesn't require any changes in the hybrid fiber/coax (HFC) network.

Bandwidth is allocated to each subscriber in a deterministic manner in dedicated Internet Protocol by using a round-robin serving scheduling mechanism within each frequency channel based on upstream time division multiple access (TDMA) sub-channels (shown in Figure 4.1) below and downstream time division multiplexed (TDM) sub-channels. Each subscriber is allocated one or more TDM channels in the downstream providing 5 to 40 Mb/s service scaleable in 5Mbs increments, and one or more TDMA channels in the upstream providing 500Kbs to 8Mbs service scaleable in 500Kbs. This is because TDMA and TDM channels are not oversubscribed to multiple subscribers, and the dedicated IP architecture completely eliminates subscriber-based contention on the access link, thereby inherently providing low latency, and low jitter connectivity to support voice applications.

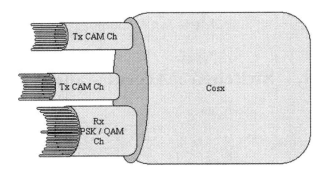

Figure 4.1: Dedicated IP Channels

Dedicated Internet Protocol is supported by two key network elements namely:

1. A head-end access router (HAR), deployed at the head-end or distribution hub, provides standard packet network interfaces like Gigabit Ethernet to the WAN, and F-type interfaces to the HFC combining and splitting networks. The HAR behaves as an IP router supporting IP forwarding between network interfaces to the WAN or MAN and subscriber-facing dedicated IP channels over the HFC access network.

2. An access gateway (AGW), deployed at the customer premise, provides an F-type interface to the HFC network and standard LAN interfaces such as 10/100Base-T Ethernet to customer premise equipment such as a subscriber telephony gateway (STG). The AGW may provide one or more RJ-11 analog voice ports that directly interface analog handsets.

Since dedicated Internet Protocol is purely an access technology for HFC, the operator is not inhibited to support any particular VoIP architecture for primary or secondary line voice, resulting in the use of dedicated IP for supporting VoIP architectures based on any combination of session initiation protocol (SIP), H.323, Megaco or media gateway control protocol (MGCP). However, since one of the established aims of the cable industry is to provide a VoIP migration strategy for cable operators with significant investments in legacy circuit-switched telephony equipment, a general idea of both a near-term and longer-term solution for IP telephony over dedicated IP is required to be dealt with later on.

In brief, dedicated Internet Protocol (IP) helps cable operators in delivering voice services, which cannot be easily provided with legacy, shared bandwidth cable modems, and establishes a migration path from TDM-based to IP-based telephony services. Cable operators can also deliver video and data services over dedicated IP without disrupting legacy cable services. Bundling voice with video and data offers the opportunity to substantially increase revenue per customer and reduce churn. Additionally, Cable operators enjoy the flexibility of providing video, data and voice services to lucrative generating small to medium enterprise (SME) subscribers via dedicated Internet Protocol. Providing a coax drop to the business location existing near the cable plant is relatively cheaper, as many businesses are located in the close proximity of a cable plant.

Cable IP telephony over the dedicated IP architecture helps fulfill the promise of advanced services that enable cable operators to offer differentiated service offerings to subscribers. Not only is IP voice highly scalable, lowering the cost of delivering voice services, it also reinforces the opportunity to provide business-class services that are immediate revenue generators.

Switched Access
It's a non-dedicated local access between the premise of a customer and the serving wire center that is interconnected to the point-of-presence of the company for service origination or termination. A switching technology similar to what people find in conventional Ethernet switches is put into practice by switched access points. Multiple wireless clients are able to communicate with the same access point simultaneously by these switches that eliminate any collisions among packets. Thus the better range and throughput are the resultant output. Switches can simultaneously send different signals in different directions, as they use phased-array antennas. This enables collision free transmission simultaneously between clients attached with the same access point.

Virtual Circuits

When a logical circuit is created within a shared network between two devices, it is called a virtual circuit. Frame Relay provides link layer communication to connection-oriented data and a defined communication exists between each pair of devices. These connections do have an association with a connection identifier. This service is implemented by using a Frame Relay virtual circuit that is a logical connection created between two data terminal equipment (DTE) devices across a Frame Relay packet-switched network (PSN).

Virtual circuits provide two-directional communication paths from one DTE device to another, and are uniquely identified by a data-link connection identifier (DLCI). Its capability of getting multiplexed into a single physical circuit for transmission across the network can often reduce the complexity of equipment and network that is required to connect multiple DTE devices.

Virtual circuits are compatible to pass through any number of intermediate DCE switches located within the Frame Relay packet-switched network.

Frame Relay virtual circuits can be categorized into two parts:

1. ***Switched Virtual Circuits (SVC's):*** Switched Virtual Circuits are dynamically established on demand and it is terminated at the completion of transmission. Communication over an SVC consists of three phases: data transfer, circuit establishment, and circuit termination. The establishment phase incorporates creation of the virtual circuit between the source and destination devices. Data transfer incorporates transmission of data between the devices over the virtual circuit. The circuit termination phase engrosses tearing down of the virtual circuit existing between the source and destination devices. SVC's are used in situations in which data transmission between devices is sporadic, largely because SVC's increase bandwidth used due to the circuit

establishment and termination phases, but they decrease the cost associated with constant virtual circuit availability.

2. ***Permanent Virtual Circuits (PVC's):*** Permanent Virtual Circuits, as is reflected by its name itself, is a permanently established virtual circuit that consists of only a single mode called data transfer. PVC's are used in those circumstances where data transfer is constant between devices. PVC's decrease the bandwidth use associated with the establishment and termination of virtual circuits, but they increase costs due to constant virtual circuit availability. The service provider generally configures PVC's when an order is placed for service.

Virtual Private Networks (VPN's)

The term VPN is used for many different services, including voice over IP (VoIP), data, remote access, and fax. A Virtual Private Network (VPN) uses the Internet and other network services as Wide Area Network (WAN) supporting strength. In this network, the dial-up connections and Leased line/Frame relay connections are replaced by local connections to an Internet service provider (ISP) or other service provider's point of presence (POP). Through a VPN a private intranet gets extended firmly across the Internet and other networks, which results in securing e-commerce and extranet connections with customers, suppliers and business partners. A Virtual Private Network is broadly categorized into three types, namely Intranet VPN's, Extranet VPN's, and Remote access VPN's.

A brief sketch of these three is given below:

- Intranet Virtual Private Networks allow private networks to be extended across the Internet or other public network service in a secure way. Intranet VPN's are sometimes referred to as site-to-site or LAN-to-LAN VPN's.

- Extranet Virtual Private Networks allow secure connections with business partners, suppliers and customers for the purpose of e-commerce. Extranet VPN's are an extension of intranet VPN's with the addition of firewalls to protect the internal network.

• Remote access Virtual Private Networks allows individual dial-up users to connect to a central site across the Internet or other public network services in a secure way. Remote access Virtual Private Networks are sometimes referred to as dial VPN's.

The operating process of these types of Virtual Private Networks is demonstrated in the following illustration:

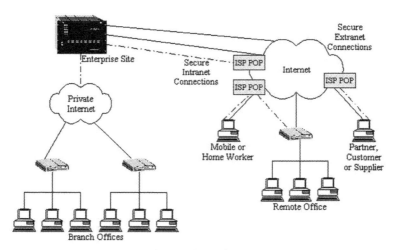

A Virtual Private Network

The objective of these Virtual Private Networks is to provide the performance, dependability, service quality, and security of conventional Wide Area Network environments that uses economical and flexible ISP or any other connections. VPN technology can also be used within an intranet to provide control access or security to sensitive information, resources or systems. As an example, VPN technology may be used to limit access to financial systems from certain users or to ensure sensitive or confidential information is sent in a secure way.

Some of the more common definitions of a VPN, out of the many that are available are as follows:

"IP tunnels between a remote user and a corporate firewall with tunnel creation and deletion controlled by the user's computer and the firewall."

"IP tunnels between an Internet service provider and a corporate firewall with tunnel creation and deletion controlled by the ISP. IP tunnels between sites over the public Internet, or over a service provider's IP network that is separate from the public Internet ISDN, Frame Relay or ATM connections between sites with ISDN B channels, PVC's or SVC's used to separate traffic from other users."

Virtual Private Networks Based on IP Tunnels

IP tunnels based Virtual Private Networks (VPN's) put a data packet in a nutshell within a normal IP packet in order to forward it over an IP-based network. The summarized packet doesn't need to be Internet Protocol, and could in fact be any protocol such as SNA, AppleTalk, DECnet or IPX. The encapsulated packet doesn't require authentication and encryption. However, encryption is used to ensure authentication and privacy with most IP based VPN's, specifically those VPN's that run over the public Internet, in order to ensure integrity & reliability of data. IP tunnels based VPN's are principally self deployed, which suggests that users buy connections from ISP's and install VPN equipment that they could configure and manage themselves, keeping aside ISP's only for the physical connections. ISP's, service providers and other carriers also provide IP tunnels based VPN services that are fully managed services with options like Service Level Agreements (SLA's) to ensure Quality of Service (QoS).

The following illustration depicts an Internet-based VPN, which uses secure IP tunnels to connect remote devices and clients.

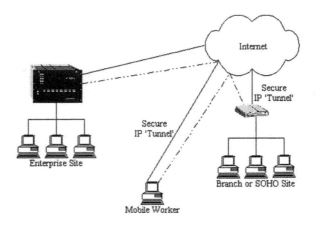

Internet - Based VPN

The four main benefits of IP tunnels based VPN's are as follows:

1. External Internet access, internal intranet and extranet access can be provided using a single secure connection.

2. Greater flexibility in deploying mobile computing, telecommuting and branch office networking.

3. Reduced telecom costs, as dedicated and long distance connections are replaced with local connections.

4. Easier e-commerce and extranet connections with business partners, suppliers and customers.

The main disadvantage of IP tunnels based VPN's is that QoS levels may be unpredictable, unreliable and erratic, as they are not yet considered as potentially high alternative solutions. Also, for VPN's based on the public Internet, higher levels of security such as authentication and data encryption are essential to ensure integrity and security of data. Note that ISP connections used for VPN's do not necessarily need to be protected by a firewall as data is protected through tunneling, encryption, etc. Also, you can use separate ISP connections for general Internet access and VPN access, or you can use a single connection with a common router, a VPN device and a firewall in parallel behind it. In some cases devices that integrate one or more of these functions can also be used.

Virtual Private Networks Based on Frame Relay, ISDN or ATM

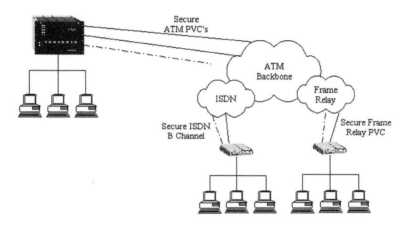

Carrier - Based VPN

Frame Relay, ISDN or ATM connections-based VPN's are exceptionally different from IP tunnels based VPN's. This type of VPN uses public switched data network services ISDN B channels, such as PVC's or SVC's for separating traffic from other users. Single or multiple B channels like PVC's or SVC's may be used between sites having additional features like backup and bandwidth on demand. Data packets neither require being Internet Protocol, nor do they need to be encrypted. However, due to widespread awareness on security issues, many users now opt for encrypting their data. The diagram given alongside shows a carrier-based Virtual Private Network that uses Frame Relay PVC's and ISDN B channels to connect devices and remote clients.

Service providers and other carriers typically provide VPN's based on public switched data networks, which is not always sure to provide fully supervised services. Additional services like Quality of Service options do exist most of the time. Europe in particular is the popular ground for this type of VPN, as the public switched data networks (PSDN) are widely available there and the Internet is less used for business purposes. There are various benefits of VPN's based on ISDN, Frame Relay or ATM connections. Some of the most important ones are as follows:

- Since data is generally carried over the private network of service providers or carriers, the concern for security is obviously less.

- Since these services are well established, loads of billing and accounting information is available.

- Despite being expensive, it is comparatively easy to get International connections, particularly for Frame Relay.

- Any type of communication, be it PBX connections or videoconferences to private data, can use these connections.

ISDN, Frame Relay and ATM services are not as widely available as ISP services and may be at the same time too expensive, which is why these types of VPN's are also considered disadvantageous. Additionally, with this type of VPN most of the time it is harder to provide extranet and e-commerce connections to customers, suppliers and business partners.

Benefits of Virtual Private Networks

One can save considerably on cost with Virtual Private Networks (VPN's) over traditional solutions. Since long-distance connections are replaced with local connections to an ISP's point of presence (POP), or local connections to a service provider or carrier network, the cost of VPN's is being considerably reduced as compared to conventional leased line, Frame Relay or other services. The two key benefits of VPN's are briefly discussed below:

1. **Cost Reduction:** With Virtual Private Networks, network managers are able reduce the overall operational cost of WAN's through reduced telecom costs. The savings can be even more in the case of a managed VPN service, as fewer networking staff is needed to manage the security aspects of

the VPN, with the ISP or other service providers managing the WAN equipment. In many cases, implementing a VPN also means that more use is made of an existing dedicated Internet connection.

2. **Flexibility:** IP tunnels-based VPN's, specifically Internet-based VPN's, offer greater flexibility with the deployment of telecommuting, mobile computing and branch office networking. Many corporations are experiencing ballistic demand increase for these services. Virtual Private Networks provide a low-cost and secure method of linking these sites into the enterprise network. It is possible to link even the most remote users into the network, as the nature of ISP services is omnipresent.

Internet Virtual Private Networks

Internet-based Virtual Private Networks are increasingly becoming widely available for its use as an alternative for dial-up remote access. By and large, people unconditionally consider VPN's as an Internet-based network providing an alternative to a public network services-based private network like Frame Relay or T1 leased lines. The Internet and Internet service providers (ISP's) have become so ubiquitous and abundant respectively, that obtaining connections in all locations has become possible, except the most remote of sites. Although some countries are still continuing with restricted access, in most countries around the world ISP's are offering connections to the Internet. So it is possible for large and small organizations both to consider the Internet for internal communications using a VPN, not just for external communication with customers, suppliers, and business partners.

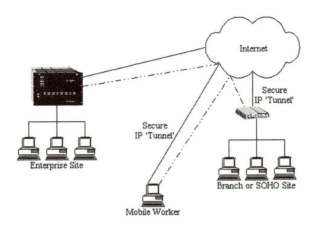

Internet - Based VPN

Internet-based VPN's can affect better flexibility and noteworthy cost savings, on account of outsourcing remote access. Modem racks, remote access servers and the other equipment necessary to service the needs of remote and mobile users can be replaced with a managed service provided by an ISP). Although Internet-based VPN's are appropriate for remote access needs, some problems still remain to be conquered prior to moving towards a full intranet VPN solution. Despite the fact that most VPN products now offer adequate levels of security, the issue of Quality of Service and Service Level Agreements still exists. While most VPN service providers can offer guarantees for connectivity and uptime, few can offer adequate throughput and latency guarantees. In addition, there are few agreements between ISP's, so unless you can use a single ISP's IP backbone for all of your connections, you are likely to suffer service degradation where connections cross boundaries between ISP's. Most users would not like to surrender the levels of Frame Relay leased lines, or ATM networks' service for anything inferior. However, it is assumed that these problems will be overcome, and Internet-based VPN's will become much more widespread for intranet as well as remote access in the years to come. In a few years global VPN services based on the Internet will become as cost-effective and as highly available as global Frame Relay and other public network services.

Public Network Virtual Private Networks

Public networks like Frame Relay, ISDN and ATM can also carry mixed data types including data, video and voice. It can also be used to provide VPN services for separating traffic and other users, with the use of Permanent Virtual Circuits (PVC's), B channels or Switched Virtual Circuits (SVC's). Encryption and authentication can optionally be used for guaranteeing the identity of users and the integrity of data. With the use of B channels, SVC's or PVC's it becomes easier to provide additional bandwidth or backup at the time of need. The traffic shaping capabilities of Frame Relay and ATM can be used to provide different levels of QoS, and because these services are based on usage, there is significant opportunity to reduce telecom costs even further by using bandwidth optimization features.

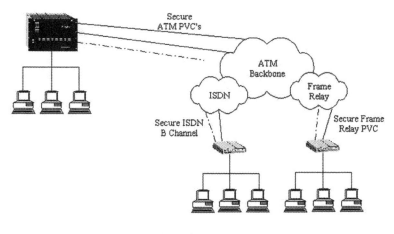

Carrier - Based VPN

Frame Relay, which is also suitable for VPN's, has particularly become a popular, widespread and low-cost networking technology. Running VPN's over a Frame Relay network allows expensive dedicated leased lines to be replaced, and makes use of the acknowledged strengths of Frame Relay, including support for variable data rates for traffic that is inclined to 'burst', bandwidth on demand, switched as well as permanent virtual circuits for connectivity on a per-call basis. The ability of Frame Relay to handle built-in buffering and disintegrated traffic helps it to use the available bandwidth optimally, which is important in a VPN environment where latency and performance are important concerns. The Frame Relays can create VPN in the following two ways:

1. **Creating a mesh of Frame Relay connections between sites:** These connections are essentially point-to-point links and are similar in concept to dedicated leased lines. Data is kept separate from other Frame Relay users as each connection uses a separate virtual circuit.

2. **Using IP tunnels over Frame Relay connections between sites:** As above, these connections are essentially point-to-point links similar in concept to dedicated leased lines, and each connection uses a separate virtual circuit. However, several separate IP tunnels can be run over each connection, and each tunnel can be encrypted and authenticated to provide additional security.

Frame Relay is an end-to-end protocol that can be run over a variety of access technologies, such as ISDN, DSL (Digital Subscriber Loop), and POTS dial-up lines. New access methods such as switched virtual circuits (SVC's), ISDN access, and backup mean that Frame Relay is now a much more reliable and cost-effective solution. Frame Relay can also run over, and interoperate with, ATM backbones, making it one of the most widely available public data networking services worldwide. As a result, major service providers and carriers have created global Frame Relay networks, which are cost-effective and offer high availability. These characteristics, coupled with authentication and tunneling encryption, make Frame Relay an ideal solution for global VPN services.

Remote Access Virtual Private Networks

Remote Access VPN's are rapidly replacing traditional remote access solutions, as they provide more flexibility and attract less cost. Remote access is the ability to connect to a network from a distant location, so that a remote access client system connects to a network access device, such as a network server or access concentrator. The client system becomes a host on the network when logged in. Some typical remote access clients may be:

- PC's with modems or ISDN connections for use at home by telecommuters.

- Laptop computers with modems for traveling users.

- Laptop computers on a shared Local Area Network, e.g. hotel chains offering LAN connection points in suites, so that Ethernet cards can be used without a modem card.

Remote access connections can be separated into two segments such as local dial, and long-distance dial. For traditional remote access and private networks, local-area users use a variety of telecommunication data services for establishing connections. Remote access long-distance users rarely have a choice other than modem access over telephone networks. The aggregation devices that the clients connect to typically use channelized leased line and primary-rate ISDN, offering dedicated circuit switched access. With VPN's local area users typically have a wider range of data services to choose from, regardless of the support at the enterprise or central site VPN equipment. However, modem access currently facilitates long-distance connections. The VPN equipment will use high-speed leased lines to the nearest POP of the chosen VPN carrier and all remote access traffic can be aggregated or routed as IP data-grams over this single link.

Advantages of Remote Access VPN's over Traditional Direct-Dial Remote Access

There are several major and minor advantages of choosing Remote Access VPN's over Traditional Direct-Dial Remote Access. Several obvious and key advantages are comprehensively listed below:

1. **Better data rates for modems:** Occurs because long-distance VPN users can dial a local modem at the VPN carrier's office, and the data rate achieved by the modem should be better than for a long-distance or international direct call.

Again, partnering with a VPN carrier to provide a service is important. For example, international VPN throughput can deteriorate badly when using the Internet as a carrier.

2. **Cost-affective dial-service for long-distance users:** When a company partners with a VPN carrier to provide global remote access, the employees are issued information on the local telephone number access points in each country for which they have support. Since local calls are significantly cheaper than national and international calls, this would appear to offer a sizable savings. These savings do of course depend on the throughput achieved and the relative cost of local, national and international calls. In most regions of the world local calls are not free, and this may mean that real savings are not achieved. For example, if local calls offer a 50% saving over national calls, but the VPN throughput means that it takes twice as long to copy mail from a central office than it would have using a direct-dial call, no telecommunication savings have been made and the company's time has been wasted. For local users with telephone lines (or ISDN), a VPN offers no dial-in cost savings and a worse service solution for the user.

3. **Less upgrading of equipment at an enterprise or central site:** With the improvement of modem technology and growth of availability of new local loop services, new hardware would be needed at a Modem Pool site. This problem is effectively handled and paid for by the VPN carriers.

4. **Scalability:** Adding hundreds of users to an enterprise Security Gateway, which only deals with IP data-grams over a high-speed leased line, gives fewer problems than adding the same number of users to a modem pool.

5. **Superior local access services:** with a conventional direct-dial remote access network, the aggregation device supported by data services dictates the data services, which can be used by the remote users. With a VPN the user can choose the best local loop service available, for example, cable modems

or DSL. This advantage is only a reality for home workers currently, but may eventually apply to mobile users.

6. **The link being used for Business Companies and for private users:** If the Internet is used as a carrier for connecting a small office or home-based office to a central site, it is possible to use the link for company and private business. It is also possible to send external mail using the ISP's mail servers and other features (e.g. fax, voice-mail, DNS, direct browsing) without burdening the company-owned servers. However, this does have the downside of billing and security issues.

7. **Better utilization of bandwidth at the enterprise or central site:** A fixed bandwidth is typically allocated to each user in the traditional approach, e.g., a 56Kb channel on a T1 circuit or an ISDN B-channel. Most remote working sessions have very low overall utilization of the reserved bandwidth allocated. Also, with a circuit switched approach, there are a fixed number of users who can be supported before new users are completely blocked. With a VPN approach, it is possible to fully utilize the available bandwidth. As the number of connected users increases, the service to each user gradually decreases, but is not completely blocked. Fully equipped users enjoying high-speed local access services may easily take advantage of any spare capacity too.

Disadvantages of Remote Access VPN's

Besides the above-mentioned advantages, Remote Access VPN's are also attached with some disadvantages that are discussed below. Most of the disadvantages, which we are going to discuss, refer to Internet-based VPN's, whereby the solutions will be available on VPN-focused carriers. Some possible disadvantages of VPN remote access are as follows:

- **Bandwidth Reservation or Quality of Service at the Enterprise or Central Site:** Bandwidth reservation is the ability to reserve transmission bandwidth on a network connection for particular types of traffic. It is much harder to achieve with VPN's than traditional networks. Some reservation can be done on out-bound traffic, but for inbound reservation to be achieved, the VPN carrier would need to help. Some inbound flow control is available with L2TP. However, controlling incoming data from power users is a problem that requires some way to flow-control input from each remote client.

- **Security:** VPN connections are made by first connecting to a POP of the public network, and then using that network to reach a remote peer to precipitate a private tunnel. Once the connection has been made to the POP, unsolicited data from other users of the public network can be received. This prescribes comprehensive and complex security measures before getting exposed to possible attacks.

- **Quality of Service:** VPN links over publicly routed networks don't normally offer any end-to-end throughput guarantees, contrary to circuit-switched or leased line data services. In addition, packet loss is variable and can be very high, as well packets can be delivered out-of-order, fragmented, and because of these QoS issues, data compression performance over a tunnel can be poor with as low as zero-history compression.

- **Accounting and Billing:** If costs of dial-in are being incurred on a link that is not directly connected to the company, which is supposed to pick-up the bill, monitoring the budget becomes a neat trick. However, this may be achieved with VPN carrier-based L2TP.

- **Centralized Tele-saving Control:** With Remote Access VPN's, managing cost-effective use of central dial links may no longer be possible.

- **Both-ways calling:** Small office or home-based office (SOHO) sites, which use ISDN to access a central site, directly enjoy the capabilities of two-way calling. For example, the central site can initiate the call, if the link is idle and traffic needs to flow from the central site to the remote site. In a VPN network, this is a capability missing from common ISP offerings today. Offering to pick up the dial-in costs incurred by partners and customers is also difficult. Currently L2TP does include support for these features.

- **Reconnection Time:** Using tunneling may increase the reconnection time for dial users. With the VPN carrier L2TP model, the client has to go through two authentication phases. One, on contacting the VPN carrier POP, and another on contact with the enterprise Security Gateway. Although the authentication exchange with the POP may well be trivial, the VPN database look-up can take time. For ISDN SOHO sites that wish to use cost-saving techniques, special features may be needed to cache these look-ups to allow rapid reconnects.

- **Support Issues:** Replacing direct-dial links with VPN tunnels may produce some very painful fault-locating missions. Due to the complexity of VPN carrier networks, the opportunities for 'hand-washing' are enormous.

- **Overhead:** VPN tunnels impose overhead for dial-in users: encryption algorithms may impact the performance of the user's system as there will be an increased protocol header overhead, authentication latency will increase, PPP and IP compression will perform poorly (compared to a direct link), and modem compression will not work at all.

- **Encryption:** When using encryption to protect a tunnel, data compression is no longer achievable as encrypted data is not compressible. This means that hardware compression over a modem connection is not possible.

- **Multimedia:** Applications such as video conferencing only work acceptably over low latency links that can offer the required minimum throughput. Currently on the Internet, latency and throughput can vary alarmingly. Multi-channel data services, like ISDN and xDSL provides short-term solution to this problem, allowing the usage of data channel for VPN tunneling, and a separate voice channel for video conferencing or business telephone calls.

Intranet Virtual Private Networks

For the purpose of providing cost-effective branch office networking, and offering significant cost savings over traditional leased-line solutions, the service of Intranet VPN's can be availed. Intranet VPN's apply to several categories of sites, ranging from SOHO sites to branch sites to central and enterprise sites. Small office or home-based office (SOHO) sites can be considered as remote access users where dial services are used. However, SOHO sites generally have more than one PC and they are really small Local Area Network sites. In an intranet VPN, expensive long distance leased lines are replaced with local ISP connection to the Internet, or secure Frame Relay or ATM connections as shown in the following diagram.

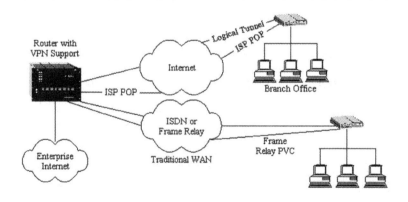

Site-to-Site VPN versus Traditional WAN

Local ISP connections can be established using many technologies, ranging from ISDN and dial-up POTS for small sites, to Frame Relay or leased lines for larger sites. New emerging Last-mile technologies like cable, DSL and wireless provide high-speed and low-cost access both at the same time. Many ISP's and service providers are now starting to support these emerging technologies for Internet access, particularly for home users and SOHO sites. The intranet market is one where traditional WAN carriers are likely to compete heavily with ISP's. Traditional WAN carriers can offer a VPN service similar to a Frame Relay service with Quality of Service (QoS) based on a Committed Information Rate. With these advantages, traditional WAN carriers are perfectly placed in providing trustworthy, secure, and low-latency intranet links by adopting their current services to support routed VPN links.

Advantages of Intranet Virtual Private Networks Solutions

There are several major and minor advantages of Intranet VPN Solutions. Some key advantages are briefly discussed below:

1. **Cost-effective Line Rental:** Typically, VPN carriers provide a leased-line feed by contracting with a traditional carrier company. Since leased lines often have a distance-related cost structure, connecting to a local POP will provide savings compared to a direct long-distance or international link.

2. **Cost-effective Backup:** If a company sticks with traditional-carrier, end-to-end data services for primary intranet links (which is advisable), the VPN carrier service may offer cheap bandwidth, backup or low-priority data routing. To do this effectively, the tunnels need the support of dynamic tunnel monitoring. For example, how does a CPE router know the effective throughput of a tunnel without an end-to-end reliable data link or intimate knowledge of the higher-layer protocol sessions carried over the link? Without a solution to this problem, path sharing between a VPN tunnel and a private leased line may give worse throughput than using just the private leased line. If the VPN tunnel is used in partnership with a private data service that had a use-based tariff, for example Frame Relay, then this solution could offer considerable savings.

3. **Scalability:** Unlike leased lines and Frame Relay PVC's, there is no additional cost for new peer-to-peer links. However, in order to offer Frame Relay-style Quality of Service, VPN

carriers may well need to introduce a per-virtual-link factor to cover costs.

4. **Cost-effective Global Virtual Backbone:** For companies that do not already have a national/international backbone, there is no cheaper option than setting up a virtual backbone using VPN carrier services.

5. **Cost-effective High Bandwidth over Last-mile:** Renting high-bandwidth leased lines, e.g., T1/E1 or T3/E3 is expensive. Cost-effective options exist for last-mile connections in some areas like wireless, xDSL, satellite and cable.

Disadvantages of Intranet Virtual Private Networks Solutions

Beside 5 advantages mentioned above, Intranet VPN Solutions are also attached with some disadvantages, out of which six possible disadvantages of intranet VPN are briefly discussed below:

1. **Packet Loss:** A VPN tunnel can sometimes suffer high packet loss and can reorder packets. Reordering can cause problems for some bridged protocols, and high packet loss may have an impact on the optimal configuration of higher-layer protocols.

2. **Denial-of-service Attacks:** Unlike a private leased line, traffic that is not from the peer remote site (tunnel end-point) can flood down the receive path of a VPN tunnel from anywhere on the public network. This unsolicited traffic may reach such a level that solicited data can no longer be retrieved. To combat this, the VPN carrier could offer to filter non-VPN traffic, or perhaps provide a bandwidth reservation or QoS service.

3. **Increased Downtime:** Decreased mean time between failures, longer lasting outages, painful problem solving and downtime compensation claims.

4. **No End-to-end Data Link in a Few Cases:** For some tunnel technologies, there is no end-to-end data link, so detection of reach ability will need to be supported at the routing layer with protocols capable of rapid failure detection and instant re-route.

5. **Aggregation of Functions:** Doing business with partners is clearly easier to achieve using the VPN model, but aggregating private tunnels, customer tunnels and web-publishing access in a single system is difficult without combined VPN and firewall capability. Separating VPN and non-VPN traffic is a sensible precaution.

6. **Latency/Delays and Multimedia:** This is very much a next-generation VPN carrier goal that will require considerable investment to do properly. There are serious doubts as to the chances of the Internet achieving success in this area in the near future. Data-link carrier companies and newly formed VPN-focus companies offering VPN services have a better chance.

Key Issues related with Virtual Private Networks

There are numbers of technological and practical issues, which requires overcoming prior to the implementation of a VPN. For a VPN to function successfully, some of these issues must provide a number of essential features, such as features that solve the problems stemming-out from private data routing across a shared public network. Some of the main features are discussed as follows:

❑ **Performance and Quality of Service (QoS)**

IP data-grams sent across the VPN carrier service may experience packet loss (silent discards) and packet reordering. Packet loss tends to be greatly increased by stated algorithms designed for point-to-point reliable links, such as PPP compression and encryption algorithms. Throughput may also vary from POP to POP, country to country, and even hour to hour. Reordering will cause problems for some LAN protocols as when running bridging over a VPN.

❑ **Security**

Since a VPN is a shared-access, routed network, security is the main area of concern. It will require the session and per-packet authentication, use of encryption, secure key exchange/re-keying, security negotiation, private address space confidentiality, complex filtering, and a host of other precautions.

❑ **Scalability**

The term scalability refers to how well a system can adapt to increased demands. A scalable network system is one that can start with just a few nodes but can easily expand to thousands of nodes. Scalability can be a very important feature because it means that you can invest in a system with confidence that you won't outgrow. If Virtual Private Network carriers are to succeed in VPN deployment, the technologies they use need to scale easily. The VPN customer will also require this at larger Security Gateway sites. Enterprises will thus require consideration of the overhead associated with security mechanisms, key management (including methods of key generation, distribution and exchange), and the overhead associated with encryption and compression, which both require a lot of processing power (Hardware compression and encryption may be needed to cope with this load).

❑ **Preventing Denial of Service Attacks**

Being connected to a public network, the VPN receive-data path can be clogged by unsolicited data to such an extent that no useful business can be achieved. Unlike a private leased line, traffic that is not from the peer remote site (tunnel end-point) can flood down the receive path of a VPN tunnel from anywhere on the public network. For client-based tunnels, there are no services currently. In the case where the VPN carrier is providing the tunnel, the VPN carrier could offer to filter non-VPN traffic, or perhaps provide a bandwidth reservation service. For the L2TP VPN carrier-based approach, the client is protected by the fact that it is not reachable via the public network, as no global address is assigned

❑ **Monitoring Actual Throughput**

In the absence of Quality of Service guarantees from the Virtual Private Networks (VPN) carriers, mechanisms are required to allow performance monitoring of tunnels.

❑ **High-Performance Routing Issues**

The nature of IP-switching filters changes, as encryptions are used from the intranet. For IP-switching/L3 switching to function on encrypted data flows, it may require the utilization of the L2TP and IPSec standards. As an example, the definition of a flow may need to make use of the IPSec protocol headers to identify a communication stream. Similarly, it may be possible to trigger on the SPI field of the ESP header used in IPSec as a means of identifying a stream. For L3 switches, which terminate secure tunnels, no fast forwarding is possible since the encrypted IP packet needs to be reconstituted prior to being forwarded. There is also the extra load of decrypting/encrypting for these secure tunnels. In time, encryption (and compression) will be present in all hosts and there will be less need for routers to terminate secure tunnels-allowing switching based on tunnel header information and requiring no encryption/decryption horsepower. Work to redefine the TOS field of IP packets as part of DiffServ may deliver the means to reinstate traffic prioritization in L3 switches for secure data flows.

❑ **Privileged Management and Supporting Tools**

Client-based software should be as transparent as possible. VPN carriers will require new management tools in order to simplify the configuration and facilitate the monitoring of a VPN corporate customer. As well, VPN customers may wish to enjoy a privileged management window into their VPN carrier-held database for the purpose of effecting changes for themselves.

❑ **Tele-Saving**

'Tele-saving' refers to the cost-effective use of WAN data services and is appropriate to all WAN links, especially for 'use-based payment' data services, such as ISDN. For clients using this type of service to access the VPN carrier network and then a tunnel server, tele-saving requires to be performed from a central site, and an Enterprise Security Gateway for data links, which are connected indirectly through the VPN carrier network. New VPN-specific 'Tele-saving' features will be required to avail the benefit of the possibility of cost effective bandwidth via a VPN link, while maintaining some layer of service with the use of more expensive private data links when desired.

❑ **Flexibility**

In order to offer a "free to go to the desired places" VPN service, VPN carriers are inclined to provide a service that can support all protocols and data links, such as PPP over anything.

Virtual Private Networks Futures

The deployment of VPN's is currently in the budding phase only. Once VPN's start enjoying wide acceptability usage, they will provide the opportunity of integrating other types of communication like multimedia and Voice over IP (VoIP) too. Although the security would remain as the primary concern for VPN's, the focus will fall more on delivering quality of service (QoS) and class of service (CoS) over IP networks as part of a VPN, once VPN products are widely available. As voice and data services merge into one (voice over IP, IP fax), new network services are being developed to offer the QoS/CoS required for fax, telephony and data. As products develop to take advantage of this opportunity, all communication devices will become IP addressable, providing voice, fax, video and data to the desktop. All of these services can make use of VPN security protocols.

Name servers will prove to be a high utility for configuring & reconfiguring VPN's. If the routers in a complex intranet VPN network were to make use of 'name servers' to locate peer routers, then these networks would have been reconfigured simply by changing the name-to-address mapping. Work is in progress to broaden the use of DNS servers for the purpose of providing a secure mechanism, based on IP Security, for routers and clients to find peer routers and servers respectively.

Next Generation Virtual Private Networks Carriers

New VPN carriers are coming to the forefront in order to take advantage of new markets. Traditional telecommunications providers alongside see that the aggregation possible with routed networks makes good sense for remote access data, since the strain reduces, as well on long-haul dial services by new VPN carriers. New 'last-mile' technologies, such as Digital Subscriber Loop (DSL) serve as a means to the telephone companies for the purpose of providing high bandwidth IP access over existing cabling (twisted-pair copper). Cable companies also offer the potential to deliver high bandwidth IP access over existing and new cable infrastructure. With the growing familiarity with delivering IP services, these new last-mile technologies places phone and cable companies in a sound position to acquire a significant share of the Internet access and VPN markets.

The main objective of the new providers is to provide VPN services. A popular technique for the purpose is building a Frame Relay or ATM backbone, is to offer VPN links with guarantees on throughput and latency, enabling customers to outsource site-to-site, remote access, and interoffice fax and voice. These networks are well placed to offer everything from voice, to site-to-site by making use of the quality of service options inherent in ATM and Frame Relay networks. To offer global services to a VPN customer with global data needs, consortiums of VPN carriers are forming to offer a uniform service internationally. Although new IP based services are becoming available, a lot of these services are based on Frame Relay and ATM.

Virtual Private Networks and Voice/Data Convergence

For providing voice, data and Internet connectivity needs, companies today are using different communication infrastructures. Components used for voice transmission include a PABX, key system or Centrex service with features such as voice mail and automated attendant. Linking voice capabilities with data applications, such as Computer Telephony Integration (CTI) may also be used. On the data side, a stackable or chassis based hub with multiple 10/100 Ethernet segments characteristically provide LAN infrastructures and WAN connectivity, which is typically provided by a router using Frame Relay or leased lines, with Internet connections for e-mail and web browsing provided through a separate firewall connection.

Companies that use a variety of data and voice services to meet their communication needs will find new alternatives becoming available that offer direct and indirect cost savings. New customer-premise routers are now appearing, which act as both Security Gateways and Multimedia Gateways. These Multi-service Routers integrate a number of LAN and WAN capabilities like hub and routing functions, and also support new applications like Internet access (browsing, publishing, e-mail, e-commerce), IP-fax, Voice Over IP (VoIP), as well as VPN traffic over a single local-loop link to a service provider POP.

The starting point for a company, which targets the utilization and advantages of VPN services, would be an initial investment in web access and web publishing. For the smooth transition from web publishing and e-mail to full e-commerce, companies should follow the below given steps:

	Activity	Instructions & recommendations
Step 1	Web publishing	Companies are already becoming familiar with accessing and publishing information and exchanging e-mail over a public routed network.
Step 2	Private remote access via VPN carrier networks: out-sourced remote access	Providing a more scalable remote access solution with cheaper access to corporate networks. The existing "modem pool" may be preserved for backup.
Step 3	Partnership access with customers, partners and suppliers (extranets)	Rather than arrange for one-off solutions each time a new partner needs to be linked to the corporate network, VPN networks provide a common technology to reduce the complexity and expense of adding new partner network links.
Step 4	An intranet VPN based on carrier networks with outsourced backbone links	Once VPN networks can offer QoS guarantees, corporate backbone links could be outsourced to manage routed networks. These would have built-in failure recovery, and should have a lower cost per month than traditional dedicated leased bandwidth.

Step 5	Full electronic commerce that is doing business over public networks	For example: electronic-fax, voice-over-IP and electronic ordering. With the growth in the reach and capacity of the Internet and the IP protocol suite, there is the promise of providing all common communications services over the same communications link-an IP data-gram service.

Ethernet Networking

Ethernet has been a relatively inexpensive, reasonably fast, and very popular LAN technology for several decades. Two individuals at Xerox PARC – Bob Metcalfe and D. R. Boggs – developed Ethernet beginning in 1972 and specifications based on this work appeared in IEEE 802.3 in 1980. Ethernet specifications define low-level data transmission protocols and the technology needed to support them. In the OSI model, Ethernet technology exists at the physical and data link layers (layers 1 and 2).

Ethernet is the most popular physical layer LAN technology in use today. Other LAN types include Token Ring, Fast Ethernet, Fiber Distributed Data Interface (FDDI), Asynchronous Transfer Mode (ATM) and LocalTalk. Ethernet is popular because it strikes a good balance between speed, cost and ease of installation. These benefits, combined with wide acceptance in the computer marketplace and the ability to support virtually all popular network protocols, make Ethernet an ideal networking technology for most computer users today. The Institute for Electrical and Electronic Engineers (IEEE) defines the Ethernet standard as IEEE Standard 802.3. This standard defines rules for configuring an Ethernet network, as well as specifying how elements in an Ethernet network interact with one another. By adhering to the IEEE standard, network equipment and network protocols can communicate efficiently.

A low-level network technology, Ethernet supports IP and most other higher-level protocols. Traditional Ethernet supports data transfers at the rate of 10 Megabits per second (Mbps). Over time, as the performance needs of LAN's have increased, related technologies like Fast Ethernet and Gigabit Ethernet have been developed that extend traditional Ethernet to 100 Mbps and 1000 Mbps speeds, respectively.

Traditional Ethernet

10Base5 technology is commonly referred as Thicknet. This technology, being the first incarnation of Ethernet, was used in the 1980's until the advent of 10Base2 Thinnet, which comes with more flexible cabling. At five millimeters, Thinnet is half the thickness of Thicknet. However, the most common form of traditional Ethernet is 10BaseT due to the inherent advantages of unshielded twisted pair (UTP) over coaxial cabling and its low cost compared to alternatives like fiber.

The following table lists these well-known forms of Ethernet technology. Besides the type of cable involved, another important factor in Ethernet networking is the segment length. A single uninterrupted network cable can only span a certain physical distance before its electrical characteristics are critically affected by factors such as line noise or reduced signal strength.

Name	Segment Length (Max.)	Cable
10Base5	500m / 1640 ft.	RG-8 or RG-11 coaxial
10Base2	185m / 606 ft.	RG 58 A/U or RG 58 C/U coaxial
10BaseT	100m / 328 ft.	Category 3 or better unshielded twisted pair

Several other less well-known Ethernet standards exist, including 10Base-FL, 10Base-FB, 10Base-FP for fiber optic networks, and 10Broad36 for broadband (CATV) cabling.

Fast Ethernet

In the mid-1990's, Fast Ethernet achieved its design goal of increasing the performance of traditional Ethernet while avoiding the need to completely re-cable existing networks. Fast Ethernet comes in two major varieties:

1. 100Base-T (using unshielded twisted pair cable)

2. 100Base-FX (using fiber optic cable)

For Ethernet networks that need higher transmission speeds, the Fast Ethernet standard (IEEE 802.3u) has been established. This standard raises the Ethernet speed limit from 10 Megabits per second (Mbps) to 100 Mbps with only minimal changes to the existing cable structure. There are three types of Fast Ethernet: 100BASE-TX for use with level 5 UTP cable, 100BASE-FX for use with fiber-optic cable, and 100BASE-T4, which utilizes an extra two wires for use with level 3 UTP cable. The 100BASE-TX standard has become the most popular due to its close compatibility with the 10BASE-T Ethernet standard. For the network manager the incorporation of Fast Ethernet into an existing configuration presents a host of decisions. Managers must determine the number of users in each site on the network that need the higher throughput, decide which segments of the backbone need to be reconfigured specifically for 100BASE-T and then choose the necessary hardware to connect the 100BASE-T segments with existing 10BASE-T segments. Gigabit Ethernet is a future technology that promises a migration path beyond Fast Ethernet so the next generation of networks will support even higher data transfer speeds.

The most popular of these to date is 100Base-T, which is a standard that includes 100Base-T4 (100Base-T2 cabling modified to include two additional wire pairs), 100Base-T2 (Category 3 or better UTP), and 100Base-TX (Category 5 UTP).

Token Ring

Another form of network configuration is Token Ring, which differs from Ethernet in that all messages are transferred in a unidirectional manner along the ring at all times. Data is transmitted in tokens, which are passed along the ring and viewed by each device. When a device sees a message addressed to it, that device copies the message and then marks that message as being read. As the message makes its way along the ring it eventually gets back to the sender, who now notes that the message was received by the intended device. The sender can then remove the message and free that token for use by others. Various PC vendors advocate Token Ring networks at different times and thus these types of networks have been implemented in many organizations.

Media

An important part of designing and installing an Ethernet is selecting the appropriate Ethernet medium. There are four major types of media in use today: Thickwire for 10BASE5 networks, thin coax for 10BASE2 networks, unshielded twisted pair (UTP) for 10BASE-T networks and fiber optic for 10BASE-FL or Fiber-Optic Inter-Repeater Link (FOIRL) networks. This wide variety of media reflects the evolution of Ethernet and also points to the technology's flexibility. Thickwire was one of the first cabling systems used in Ethernet but was expensive and difficult to use. This evolved to thin coax, which is easier to work with and less expensive.

The most popular wiring schemes are 10BASE-T and 100BASE-TX, which use unshielded twisted pair (UTP) cable. This is similar to telephone cable and comes in a variety of grades, with each higher grade offering better performance. 'Level 5' cable is the highest, most expensive grade, offering support for transmission rates of up to 100 Mbps. 'Level 4' and 'level 3' cables are less expensive, but cannot support the same data throughput speeds. Level 4 cable can support speeds of up to 20 Mbps, whereas level 3 up to 16 Mbps. The 100BASE-T4 standard allows for support of 100 Mbps Ethernet over level 3 cable, but at the expense of adding another pair of wires (4 pair instead of the 2 pair used for 10BASE-T). For most users, this is an awkward scheme and therefore 100BASE-T4 has seen little popularity. 'Level 2' and 'level 1' cables are not used in the design of 10BASE-T networks.

For specialized applications, fiber-optic, or 10BASE-FL, Ethernet segments are popular. Fiber-optic cable is more expensive, but it is invaluable for situations where electronic emissions and environmental hazards are a concern. Fiber-optic cable is often used in inter building applications to insulate networking equipment from electrical damage caused by lightning. Since it does not conduct electricity, fiber-optic cable can also be useful in areas where large amounts of electromagnetic interference are present, such as on a factory floor. The Ethernet standard allows for fiber-optic cable segments up to 2 kilometers long, making fiber optic Ethernet perfect for connecting nodes and buildings that are otherwise not reachable with copper media.

Topologies

A network topology is the geometric arrangement of nodes and cable links in a LAN, and is used in two general configurations, i.e. bus and star. These two topologies define how nodes are connected to one another. A node is an active device connected to the network, such as a computer or a printer. A node can also be a piece of networking equipment such as a hub, switch or a router. A bus topology consists of nodes linked together in a series with each node connected to a long cable or bus. Many nodes can tap into the bus and begin communication with all other nodes on that cable segment. A break anywhere in the cable will usually cause the entire segment to be inoperable until the break is repaired. Examples of bus topology include 10BASE2 and 10BASE5.

Fast Ethernet and 10BASE-T Ethernet use a star topology, in which a central computer controls the access. Usually one end of the segment hosts a computer, and the other end is terminated in a central location with a hub. Since UTP is often run in conjunction with telephone cabling, this central location can be a telephone closet or other area where it is convenient to connect the UTP segment to a backbone. The primary advantage of this type of network is reliability, for if one of these point-to-point segments gets broken, it will affect only the two nodes on that link. Other computer users on the network continue to operate as if that segment were nonexistent.

Collisions

As Ethernet is a shared media, there are set guidelines for sending data-packets to protect data integrity and avoid conflicts. Nodes determine the availability of the network and if it is available for sending packets. It is possible that two nodes at different locations attempt to send data at the same time. When both PC's are transferring a packet to the network at the same time, a collision will result.

Minimizing collisions is a crucial element in the design and operation of networks. Increased collisions are often the result of too many users on the network, which results in a great deal of contention for network bandwidth. This can slow the performance of the network from the user's point of view. Segmenting the network, where a network is divided into different pieces joined together logically with a bridge or switch, is one way of reducing an overcrowded network.

Ethernet Products

The discussed technology and standards help define the specific products, which is used by network managers to build Ethernet networks. The key products required for building an Ethernet LAN are discussed comprehensively below:

❏ **Transceivers**

Transceivers are the device that is used for establishing connection between nodes and the different Ethernet media. Most computers and network interface cards contain a built-in 10BASE-T or 10BASE2 transceiver, which allows them to directly establish a connection with the Ethernet without using an external transceiver. Many Ethernet devices provide an AUI connector to allow the user to connect to any media type via an external transceiver. The AUI connector consists of a 15-pin D-shell type connector that happens to be male on the transceiver side, and female on the computer side. Transceivers are also in demand for Thickwire (10BASE5) cables in order to allow connections.

A new interface named Media Independent Interface (MII) was developed for Fast Ethernet networks in order to put forward a flexible way of supporting 100 Mbps connections. The MII is a widely accepted way of connecting 100BASE-FX links and copper-based Fast Ethernet devices.

❑ **Network Interface Cards**

Network interface cards are popularly termed as NIC's, and they are used to connect a PC to a network. With the NIC a \hysical connection between the networking cable and the internal bus of a computer is established. Different computers have different bus architectures. ISA expansion slots are most commonly found on 386 and older PC's, and PCI bus master slots are commonly found on 486/Pentium PC's. NIC's come in three basic varieties, namely 8-bit, 16-bit, and 32-bit. The larger the number of bits that can be transferred to the NIC, the faster the NIC can transfer data to the network cable.

Many NIC adapters comply with Plug-n-Play specifications. On these systems, NIC's are automatically configured without user intervention, while on non-Plug-n-Play systems, and configuration is done manually through a setup program and/or 'DIP switches'. A variety of cards are available to support almost all networking standards, including the latest Fast Ethernet environment. Fast Ethernet NIC's are often 10/100 capable, and will automatically set to the appropriate speed. Full duplex networking is another option, where a dedicated connection to a switch allows a NIC to operate at twice the speed.

❑ **Hubs/Repeaters**

Two or more Ethernet segments of any media type are connected together with the help of Hubs/Repeaters). When segments in larger designs cross their maximum length, the quality of signal starts deteriorating. With the help of Hubs the required signal amplification achieved now allows a segment to travel up to a longer distance. A hub takes any incoming signal and simply repeats it out all available ports.

Ethernet hubs are compulsory in star topologies such as 10BASE-T. Several point-to-point segments are aggregated into one network by a multi-port twisted pair hub. One end of the point-to-point link is attached to the hub and the other is attached to the computer. If the hub is attached to a backbone, then all computers at the end of the twisted pair segments can communicate with all the hosts on the backbone. Ethernet rules limits the number and type of hubs in any one-collision domain.

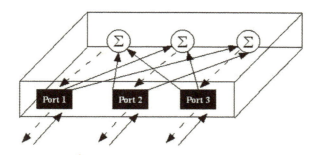

One very important fact about hubs is that they only allow users to share Ethernet. A network of hubs (or repeaters) is referred as Shared Ethernet, which means all associated members of the network are competing against each other to transmit data onto a single network i.e. collision domain. In other words, we can say that only a certain percentage of the available network bandwidth will be provided to individual members of a shared network. The number and type of hubs in any one collision domain for 10 Mbps Ethernet is limited by the following rules:

Network Type	Max Nodes Per Segment	Max Distance Per Segment
10BASE-T	2	100m
10BASE2	30	185m
10BASE5	100	500m
10BASE-FL	2	2000m

❑ **CAT5 Cable (Class 6 – Class 7)**

This is unshielded twisted pair (UTP) data grade cable and is reasonably priced. There is a big similarity between CAT5 cable and ubiquitous telephone cable; the only difference being the pairs under the former are twisted more tightly. A CAT5 cable generally consists of four twisted data pairs. Out of these four data pairs two are essentially used by Ethernet, one pair is for the transmission and the last one for receiving. The cost of this 4-pair cable is about $0.30/meter. When Fast Ethernet is carried over CAT5 cable, it is referred as 100BASE-TX.

The performance of the latest 'Class 6' or 'Class 7' cables (e.g. Belden MediaTwist 1872A) is even more enhanced than that of CAT5 cable. These latest cable types have less cross talk and loss within the cable, and are more immune to interference from outside sources.

❑ **Fiber Optic Cable**

Installation of fiber optic cabling is not as tough a science as it once used to be. In the recent years, big leaps have been registered in the cost of durability and ease of termination of this media. Fiber optic cable offers two main advantages over twisted pair cable. First, data may be carried much further over fiber, but even more important, fiber is now immune to electromagnetic interference.

Multimode and Single Mode are the two basic types of fiber that is in use today. Multimode fiber is used extensively in the data communications industry. Fast Ethernet carried over multimode fiber is referred as 100BASE-FX. Ethernet may be carried up to 2 kilometers on this fiber. The telecom industry extensively uses the single mode fiber, as it allows much greater run lengths than multimode fiber. However, numerous data-com products offer this capability, despite the fact that there is no official standard for carrying Ethernet over a single mode fiber. Two strands are required, one for transmission, and one for receiving each Ethernet link. Both multimode and single mode fiber cables are available with varying numbers of strands. The cost of 4-strand cable is about $0.80/meter.

❑ **Switch**

A switch – a multi-port device that filters and forwards data packets between devices on a network, unlike a standard repeater hub, is capable of forwarding the data packet to the correct port by reading the destination address of each packet. This attribute of intelligence in the switch helps a given device to receive only those packets that are addressed to it. Another difference between hubs and switches is the inherent ability of a switch to avoid data collisions. Should two switch ports attempt to transmit to the same port, the data enters a queue and is then transmitted serially. An Ethernet network based around switches can be wired in a star configuration with the switch at the center, or in a ring configuration using many switches. Like hubs, switches are available in varying port counts, and many offer stacking capability.

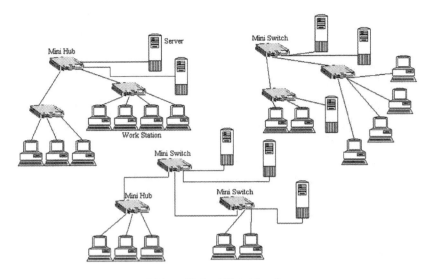

Switches and Dedicated Ethernet Examples

Switches achieve fault tolerance capabilities, which don't come with hubs. By using techniques, such as spanning tree, meshing, and trunking, switches succeed in carving some amount of recovery from failed network nodes and/or connections. Just be careful, for all of these techniques are not standardized. Therefore, the safest bet would be sticking to a single switch vendor for your design.

Ethernet in Use

Irrespective of its speed or layout, the functions of an Ethernet network are almost identical. Network devices typically possess a network interface card (NIC), such as a network adapter that interfaces directly to the system bus. The NIC possess' a cable connector such as the RJ-45 connector that is used in modern telephones. However, the original versions of Ethernet used a variety of different connectors. Data sent over the Ethernet exists in the forms of frames, which contains two headers and a data section having a combined length of up to 1518 bytes. The standard calls for frames to be broadcast to all devices seeing network adapters, which must recognize explicitly and discard all frames that are not intended to reach them.

Devices looking forward to transmit on the Ethernet first perform a very quick check-up for determining whether the medium is available, or a transmission is currently in progress. If the Ethernet is found to be present, then the device activates the transmission process. However, the Ethernet standard does not prevent multiple devices from transmitting simultaneously. These so-called collisions cause both transmissions to fail and require both devices to re-transmit at a later time. A specialized algorithm is used with Ethernet to determine the proper waiting time between re-transmissions.

As mentioned, the reach of Ethernet cables is limited, and these distances, which sometimes happen to be as short as 100 meters, are not adequate to cover medium and large-sized workgroups. A repeater in Ethernet networking is a device in which multiple cables are joined and greater distances are spanned. An Ethernet hub is one popular type of repeater device.

Network Design Criteria

There are design rules associated with Ethernets and Fast Ethernets, which are essential to be followed in order to achieve correct functioning. The electrical and mechanical design properties of each type of Ethernet and Fast Ethernet media define the number of repeaters, maximum number of nodes and maximum segment distances.

Ethernet

A repeater-based network functions with the timing constraints of Ethernet. Despite that the speed of the electrical signals on the Ethernet media is close to the speed of light, and the signal takes considerable time to travel from one end of a large Ethernet to another. A rough estimate for a signal to reach its destination, keeping in tune with the Ethernet standard, is roughly 50 microseconds. Few bulleted highlights related to Ethernet are as follows:

- Ethernet is subject to the '5-4-3' rule of repeater placement.

- The network can only have five segments connected.

- It can only use four repeaters.

- Out of the five segments, only three can have users attached to them; the other two must be inter-repeater links.

If these repeater and placement rules are violated by the design of the network, then the guidelines set for the timing will not be achieved, resulting in the sending station resending the packet. This can lead to packets lost and a disproportionate number of resent packets that can subsequently slow your network performance and create trouble for applications.

Fast Ethernet

Since less time is consumed, as compared to the regular Ethernet to transmit the minimum packet size, Fast Ethernet enjoys modified repeater rules. The length of the network links allows for a fewer number of repeaters. There are two classes of repeaters in Fast Ethernet networks namely Class I repeaters and Class II repeaters. The former have latency/delay of 0.7 microseconds or less and are limited to one repeater per network, whereas the Class II repeaters have a latency of 0.46 microseconds or less, and are limited to two repeaters per network.

The following are the distance (diameter) characteristics for these types of Fast Ethernet repeater combinations:

Fast Ethernet	Copper	Fiber
No Repeaters	100m	412m*
One Class I Repeater	200m	272m
One Class II Repeater	200m	272m
Two Class II Repeaters	205m	228m
* Full Duplex Mode 2 km		

When conditions require greater distances or an increase in the number of nodes/repeaters, then a bridge, router or switch can be used to connect multiple networks together. These devices join two or more separate networks, allowing network design criteria to be restored. Switches allow network designers to build large networks that function well. The impact of repeater rules on network design gets reduced with the reduction in costs of bridges and switches. Connection between networks using one of these devices is referred to as a separate collision domain in the overall network.

Ethernets Becoming Too Slow at Times

Performance of the Ethernet deteriorates because of the expanding user-base of a shared network, and also because applications requiring more data are added. This problem is precipitated because all users on a shared network act as competitors for the Ethernet bus. If 30 to 50 users simultaneously share a moderately loaded 10 Mbps Ethernet network, the network will only sustain throughput in the neighborhood of 2.5 Mbps after accounting for inter-packet gaps, packet overhead and collisions.

Increased number of users and packet transmissions also raise the possibility of a collision. Collisions occur when two or more nodes attempt to send information at the same time. When a collision is detected, each node shuts off for a random time before attempting another transmission. The possibility of collision increases with shared Ethernet, as more nodes are added to the shared collision domain of the shared Ethernet. One of the steps to alleviate this problem is to segment traffic with a bridge or switch. A switch can replace a hub and improve network performance. For example, an eight-port switch can support eight Ethernets, each running at a full 10 Mbps. Dedicating one or more of these switched ports to a high traffic device like a file server is another alternative. Multimedia and video applications demand as much as 1.5 Mbps of continuous bandwidth. As we have seen above, a single such user can rarely obtain this bandwidth if they share an average 10 Mbps network with 30-50 people. The video will also look disjointed if the data rate is not sustained. Ergo, greater throughput is required to support this application. Ethernet switches, when added to the network, provide a number of augmentations over shared networks, foremost being the ability to divide networks into smaller and faster segments. Ethernet switches examine each packet and determine its destination target, and then forwarding it to only those ports where the packets are destined to reach. Modern switches do all these tasks without any delay. Besides deciding when to forward and filter the packet, Ethernet switches also completely regenerate the Ethernet packet. This regeneration and re-timing allows each port on a switch to be treated as a complete Ethernet segment, capable of supporting the full length of cable along with all of the repeater restrictions.

Additionally, Ethernet switches detect bad packets and immediately drop it from any future transmission. This activity of sanitization keeps problems isolated and confined to a single segment, keeping them away from disrupting other network activities. This aspect of switching is extremely important in a network environment where hardware failures are to be anticipated. Full duplex is another method used to increase bandwidth to dedicated workstations or servers and doubles the bandwidth on a link, which provides 20 Mbps and 200 Mbps for Ethernet and Fast Ethernet respectively. If you are using full duplex, the switch is programmed to support full duplex operation, and special network interface cards are installed in the server or workstation.

The next logical step to increase performance is the implementation of Fast Ethernet. Higher traffic devices can be connected to switches or each other through 100 Mbps Fast Ethernet, in order to increase the bandwidth to a great extent. Many switches are designed keeping this in mind, and use Fast Ethernet uplinks for connection to a file server or other switches. Ultimately, Fast Ethernet can be deployed to the users' desktops by using Fast Ethernet switches, repeaters, and equipping all computers with Fast Ethernet network interface cards. Having understood the underlying technologies and products that are used in Ethernet networks, we can now march ahead to a discussion of some of the most popular real world applications.

Chapter Five
Types of Voice Communication Systems

a) **KTS**

b) **PC-Based PBX/Key System**

c) **IP Centrex**

d) **IP Telephone Systems**

e) **IP Enabled Digital Telephone System**

f) **IP Gateways**

Types of Voice Communication Systems

The importance of Internet Protocol (IP) technology revolves around how it will reform the way enterprises communicate with suppliers, customers, and employees. It is an apprehension that would normally take some years before organizations fully exploit the potential of the Internet and IP network technology, but it introduces new ways of communicating and conducting business. The successful implementation of IP will not require business users to conform to the technology, but rather, the technology will conform to the users and how they want to interact with the world.

IP technology is leading to the re-definition of business communication systems. Business communication systems from an enterprise standpoint are much more than just telephone systems. These systems are increasingly defined by the integration of voice, data, and applications all operating on a single network in a seamless manner. This is referred to as convergence and it is what allows enterprises to become more efficient by enabling customization of the user experience. IP technology is the catalyst that is causing this convergence trend to take shape.

IP-based applications will allow enterprises to transform their business in a way that will enable them to enjoy a sustainable competitive advantage. The IP-based voice application will be integrated with other applications such as Unified Messaging, ACD, CRM, and e-commerce to allow enterprises to work the way that they want to work. There will be linkages between various databases connecting remote or traveling employees to the office and their customers, all the time, from anywhere, and through various portals. True and useable mobility will be realized.

IP applications will be smart and the business communication systems will be configurable and adaptable. For example, imagine your cell phone and laptop computer connected to your enterprise network wirelessly, with the ability to know when you are in the office, which automatically 'configures' your system based on this, while also downloading updated contact information to your PDA or cell phone. The system could also route your cell phone calls to your office telephone. Information will be presented to users the way they need it based on where they are. Smart and configurable systems will create the winning advantage in the marketplace.

Some of the popular Voice Communication Systems are discussed below:

KTS

The name KTS is often given to a small PBX. A typical key telephone system (KTS) consists of a key service unit (KSU), lines, telephones/stations, line and station cards, and option cards. The KSU includes a power supply, central processing unit, trunk, extension, and optional equipment interfaces. The KSU is housed in a cabinet that can be wall-mounted, freestanding, installed in racks, and is usually equipped anywhere from 1 to 200 extensions.

PC-Based PBX/Key System

New technology continually brings new capabilities to business communication systems. This new technology comes in different forms.

- Advances and improvements in traditional PBX systems make them more functional and easier to use.

- The adaptation of data networks to carry voice provides ways to converge multiple networks together using Voice over IP (VoIP) technology.

A PC-Based PBX/Key System consists of a Telephony Server connected to the LAN. Voice cards installed in the server are connected to standard analog telephones over twisted pair wire. Additional hardware is installed in the server to support a connection to the central office trunk lines. An Ethernet card provides direct connection to the LAN for computer telephony and maintenance.

These systems do not use the data network to transmit voice. Basically, these are key systems ported to a Windows NT server.

IP-enabled PBX Systems

The rock solid reliability of PBX systems cannot be argued. They almost never go down and rugged digital telephones are equally durable. They offer almost all the telephony features anyone could need. However, while PBX systems support various Computer Telephony Integration (CTI) and IP-enabling applications, we are still converging two technologies somewhat, and the telephones operate on a separate network from the organization's data network. This requires you to maintain two separate voice and data networks. This may be an advantage or disadvantage depending upon the amount of IP network infrastructure you already have in your enterprise. IP-enabling these PBX systems provide VoIP trunk access and remote telephone user applications over IP networks to supplement access through the public switched telephone network. The IP-enabled PBX architecture typically involves the addition of IP trunk cards and IP station cards, with Ethernet interfaces to existing PBX systems as shown in the example below.

IP-PBX

The IP-PBX that operates in a pure IP environment is based upon PC server technology, and uses a single network of communication devices and wiring for both voice and data traffic. This network consolidation is assumed to result in decreased network administration, thus making deployment of services and applications easier. Hosting telephones connected through one IP network, either locally via a LAN, or remotely at any location via a private Intranet or the public Internet, provides the flexibility of distributed configurations and remote telephones users. The IP network will provide all the call switching, regardless of whether calls originate from the public switched telephone network, digital or analog telephones, or IP telephones, as shown in the example below.

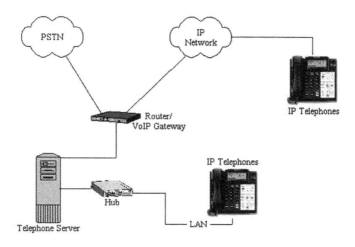

IP Centrex

IP Centrex is an emerging model for delivering telecommunications services to enterprise customers via a managed IP network. IP Centrex is a hosted application where the telephony equipment resides at the service provider from whom the enterprise client buys IP telephones and features such as ACD, voice mail, and unified messaging. The basic configuration of an IP Centrex application is an IP telephony server at the central office that provides basic call control and features such as forwarding, call waiting, and conferencing. These servers provide a gateway to the PSTN, as well as the enterprise IP Centrex gateway.

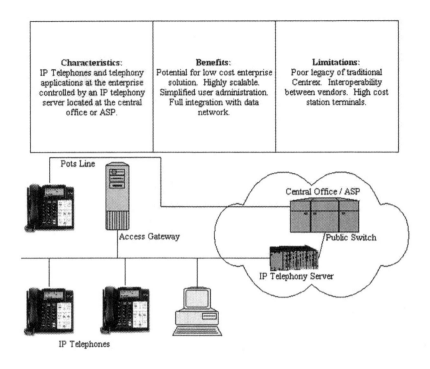

Characteristics:	Benefits:	Limitations:
IP Telephones and telephony applications at the enterprise controlled by an IP telephony server located at the central office or ASP.	Potential for low cost enterprise solution. Highly scalable. Simplified user administration. Full integration with data network.	Poor legacy of traditional Centrex. Interoperability between vendors. High cost station terminals.

Fig: IP Centrex

IP Telephone Systems

The distinguishing characteristic of an IP Telephone System is the fact that it provides PBX capabilities and delivers voice over the LAN. These switches, which can run on Windows NT, Unix, or a proprietary Operating System, control and distribute calls over the LAN. The stations are connected via Ethernet and may require an external power source. Typically, the stations communicate peer-to-peer and only require the telephony server when call control features such as conference, transfer, and voice mail access are invoked. Voice protocol may be SIP, H.323, or a proprietary algorithm.

Characteristics:	Benefits:	Limitations:
Windows NT, Unix, or proprietary telephony server, LAN-based IP telephones.	True IP system. One wire to desktop. Full integration with data network. Easy administration	Low reliability. Questionable voice quality. Limited feature set. High cost.

Fig: IP Telephone System

IP-Enabled Digital Telephone System

While the IP Telephone system is a revolutionary change to the basic voice communications architecture, the IP-Enabled Digital Telephone system is an evolution of the current model. The basic premise of the IP-Enabled Digital Telephone system is to provide LAN connectivity via an internal Ethernet connection. This Ethernet connection may then be used to pass call event data to LAN-based applications such as unified messaging systems, call accounting systems, and database applications. In addition, the Ethernet connection may also be used to send packetized voice out onto the network to an IP Station.

The benefit of this type system is two fold. First, these systems ensure the viability of the traditional enterprise Key System/PBX by providing access to all of the benefits of the IP Telephone System without sacrificing features, security or quality of service. Second, because it is a hybrid system, based on applications or user requirements, both digital telephones and IP telephones are supported. Similar to the IP telephones used with IP Telephone Systems, voice protocol for IP Telephones can be SIP, H.323 or a proprietary algorithm.

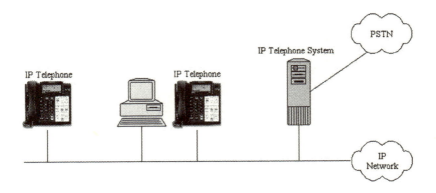

IP Gateways

IP Gateways include toll bypass products and off-premises IP extensions. The defining characteristic of an IP Gateway is that it is a point-to-point IP connection from one enterprise telephone system or station instrument to another. In addition, some IP gateways may be connected directly to the IP network to provide a reduced cost long distance option. IP Gateways accept incoming calls, packetize the voice traffic, and then send it across a packet-based network using the IP transport protocols. At the receiving end, the gateway converts the packet voice back into a TDM stream and outputs the voice in the traditional manner.

Characteristics:	Benefits:	Limitations:
IP bridge between digital or analog telephone systems and station sets.	Potential for toll cost savings. Cost effective solution for remote employees.	Subject to QOS of IP network. Requires fixed amount of bandwidth to assure quality. Dropping cost of long distance minimizes value proposition.

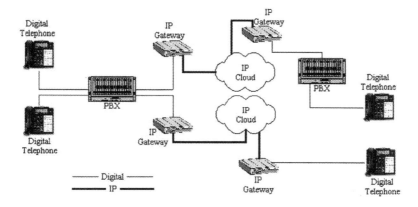

Fig: IP Gateways

Chapter Six
Applications of IP Telephony

a) **Messaging Systems**

b) **Facsimile**

c) **Voice processing**

d) **Electronic mail**

Applications of IP Telephony

The recent technological advancements of IP have provided a considerably different set of alternatives for the provisioning of long distance voice. There are three main factors that are responsible for this development.

- IP is a packet network that uses network resources only when these resources are required, and not all of the time.

- Speech compression allows for the transmission of voice in a highly compressed form while retaining the quality that the consumer demands. This compression enables the integration of speech with IP.

- Latest available technologies are now accessible, which can enable simple integration of the normal telephone network with speech compression systems and in turn with IP.

Thus, as IP universality combined with low cost, in conjunction with speech quality and ease of implementation, has now established a basis for a new and innovative market, namely long distance VoIP.

The existing model for IP voice comprises of one's own personal computer as the gateway point to access the IP and to obtain the long distance service. Worldwide research is in the process and new models are constantly being developed that can make four new key assumptions:

- To obtain universal admission the user must access the system through his/her own telephone, even if that is a rotary dial telephone.

- The access to IP must be transparent to the user, just dialing their numbers and never knowing that there is another virtual long distance carrier in place.

- There must be all of the infrastructure elements in place, such as billing and customer service to ensure that the quality of the overall offering is a first class service.

- The service must be of a sound and voice quality that is as indistinguishable from the telephone network as possible.

Worldwide IP Telephony Service Revenues

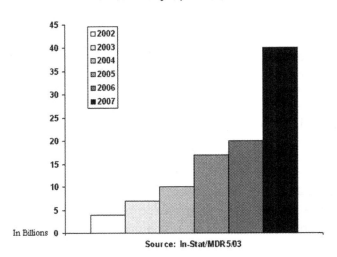

Source: In-Stat/MDR 5/03

According to In-Stat/MDR, even though there were several limitations of IP Telephony, and certainly it has had its faults in the past, the technology went through a boom and bust cycle, yet it still remains a viable consumer service. It is also believed that consumer IP Telephony is presently in the beginning of a transition phase with the growth of broadband, which will enable IP Telephony providers to merge lower rates with greater quality and features. It is also predicted that broadband providers will look at IP Telephony as a technology that will assist in the growth of broadband usage worldwide.

Overall by 2007, the US IP Telephony market is forecasted to grow to over 5 million active subscribers. Even though it shows a 5-fold increase in subscribers over 2002, it still lags US Plain Old Telephone Service (POTS) with over 100 million households.

Securing Application and IP Telephony Communications Servers

IP telephony is the collection of technologies that emulates and extends today's circuit-switched telecommunications services to operate on packet-switched data networks based on the Internet Protocol (IP). Defined in this way, IP telephony encompasses the other terms and extends those capabilities even further to include new telecommunications applications made possible by the convergence of voice and data. The terms IP Telephony, Internet Telephony, Voice-over-IP, and Fax-over-IP are all related and are often used as synonyms. The backbone of the IP telephony systems are the Communication Servers, which could be used as standalone, or can be integrated as in the case of IP-enabled PBX's or new office-in-a-box solutions. Equally important are application servers delivering contact center, multimedia applications, and unified messaging.

Telecommunications vendors have designed highly robust communications networks, using proprietary or commercially available real-time OSes and UNIX. OSes are considered to be secure while UNIX is considered reasonably robust. However, the most common OS in the data world is, not surprisingly, also used extensively for application servers supporting IP telephony and communication intensive applications called Windows NT. These use a hardened version of Windows NT with off-the-shelf security software for functions such as anti-virus protection, intrusion detection, and login audits. Hardening Windows NT starts with the requirements that server cloning should be avoided and that the media from which the operating system is downloaded must be trustworthy.

From a management perspective, a physically dedicated Ethernet port configured as on a virtual LAN (VLAN) should be set up, with all non-management traffic blocked at the routing level via access lists and firewalls. Off-net access for suppliers, system integrators, and/or VAR's can be provided via IP VPN's. Unused ports (e.g., for consoles or remote modem access) should be turned-off. Only authorized application software should be run on these servers. Multiple levels of privileges (monitor, configure, control) should be supported for authenticated operational personnel. User passwords must be securely stored, password formatting and change management strictly controlled. Management traffic (such as billing information) can be optionally encrypted even for internal transmission, again through IP VPN technology.

The shift to IP telephony offers higher efficiencies in the transport of voice and data, and as a result, lower telecommunications costs to end users. Moreover, as IP telephony is gradually evolving, the days are not very far when VoIP will be able to match all the functionalities of voice communications, which are currently supported by the PSTN. However, interoperability among the IP telephony products of different vendors is the first major hurdle to overcome.

However, the real promise of IP telephony will be realized with the next wave of advanced services that will begin to surpass the capabilities of the PSTN. Such future services include:

- **Multimedia Conferencing** – This service will allow multiple users to communicate by voice and video.

- **Multicasting** – It has the ability to broadcast voice, video, and/or data to a large targeted group.

- **Collaborative Workgroup Applications** – These applications can facilitate multiple users to interact verbally and visually while sharing access to common data and applications.

- **Call Center Applications** - Using IP telephony to enhance today's Web-based E-commerce through interaction with live service representatives.

- **Unified Messaging** - IP telephony provides a perfect medium for 'unified messaging' systems, which combine voice mail, e-mail and faxes into one easy to access platform.

- **IP Call Waiting** - Expanding the number and type of simultaneous incoming calls beyond the two voice calls that the PSTN currently supports.

IP telephony and the future ubiquity of IP-based networks will result in successive waves of innovation in technology, in devices, and in applications.

Messaging Systems

VoIP is becoming gradually more accepted as enterprises see dual benefits of reduced operational costs from transferring inter-company voice traffic to their corporate data networks and enhanced productivity through converged real-time voice/data networks and messaging systems.

VOIP can enable unified messaging by providing a single point of access to all three message-types, i.e. voice, fax and e-mail from virtually any communications device, such as the telephone, personal computer or Web browser through the Internet.

Enterprises adopting VOIP systems are using unified messaging applications over their newly converged networks. These enterprises are realizing that even if UM doesn't reduce management costs, a migration to VOIP could make UM more appealing than it was in the TDM world, and for one simple reason – IP-based unified messaging systems are likely to be less costly to purchase and integrate.

Voice Messaging over IP allows service provider subscribers to check and access messages from any phone and to perform the following tasks:

- Create multiple personalized greetings programmed to play at different times, including times when the line is busy, when there is no answer, and when calls are received after the close of business.

- Place a new call or respond to a message without leaving the messaging system, which is known as the 'Return Call' feature. This feature allows subscribers to respond to the message, forward it to someone else, or place a new call and return to the messaging system to continue processing other messages. All messages and calls can be handled with a single call.

- Leave messages for multiple recipients with a single call.

- Designate or prioritize messages so that subscribers can retrieve messages based on priorities.

- Locate a subscriber mailbox by name or telephone number.

- Forward voice messages as e-mail attachments to any e-mail user, enabling users of different voice-mail systems to share voice-mail messages.

- Receive message-waiting indication by pager, stutter dial tone or an indicator light on a telephone.

Facsimile

Fax messaging over IP allows subscribers to receive faxes anywhere by redirecting fax messages from his/her UM mailbox to a nearby fax machine. Fax messaging over IP also enables subscribers to perform the following tasks:

- Determine by using their telephone, which faxes have arrived, the arrival time, and the identity of the sender.

- View faxes as .tif files from an e-mail client and save them in separate folders.

- Forward fax messages to other people as e-mail attachments.

- Receive immediate paging notification when fax messages arrive.

- Have greater privacy by printing faxes from their mailboxes when they are ready to view them.

Voice Processing

In voice processing applications, a VOIP trunking unit or 'host' is used to replace a Key system unit, which handles all of the incoming telephone lines on one side and all of the extension lines for the individual phones on the other side. The host takes care of deciphering digital packets to analog signals for the outside phone trunks, and vice-versa digitizes all of the inbound analog calls to data packets for distribution across the LAN. All voice signals on the 'inside' run on the LAN Ethernet, and phones can be either personal computers or actual phone handsets.

Electronic Mail

E-mail messaging is one of the foremost applications used over IP and allows its users to access e-mail messages from a phone. Some of the major facilities provided by this form of messaging are:

- It is used to identify voice, e-mail, and fax messages in an e-mail inbox, which saves time by using one access device for all messages. Voice messages/attachments can be further played as streaming audio or .wav files.

- Listening to e-mail messages from a telephone using the text to speech (TTS) feature.

- Responding to an e-mail message over the phone with an audio attachment.

- Receiving a paging notification on the arrival of e-mail messages.

- E-mail messaging over IP supports both Point of Presence (POP) and Internet Messaging Access Protocol (IMAP) clients.

Chapter Seven
Design Considerations

---

a) **Networking**

b) **Hardware, Software & Protocol Requirements**

c) **Integrating with Legacy Systems**

Design Considerations

During the past 15 years data networking has moved from a hierarchical design to a distributed network. In the present scenario, it is considered to be the most viable solution to provide a reasonable approach for communication. Today's data networks provide two main functions:

a) Physically connecting to the end user (PC) or the WAN links.

b) Switching the data traffic between them.

Some of the major applications of data networks provide facilities with add-ons, such as: e-mail, spreadsheets, video, and so on.

For example, in the data-networking arena, how we have moved from large and exclusive IBM SNA networks to inexpensive routed networks that connect everyone, potentially, with everyone else. Unpicking the building blocks that make up the network and providing something with which everyone can participate, is how this was created. The same is true of voice systems. All we need to do is look at how we unpick the building blocks and review the technical considerations.

Call Management

Call management (or call control) and directory services are key elements in the network design. Whether this is built into the software on a PBX or a Soft PBX, this function is critical. It also defines the capability of the network in terms of availability, performance, scalability and administration. Taken as a fundamental building block, its performance as a switching engine is paramount.

Voice Quality

Originally, IP was conceived as a transport protocol for data traffic, not real-time voice and video traffic. As a result, when IP networks are heavily loaded, real-time delivery of voice is affected by serious quality issues due to the lack of provisioning for voice traffic. These issues include delay, jitter, and packet loss.

As discussed earlier, delay is measured by the amount of time it takes for a packet to travel from point 'A' to point 'B'. High end-to-end delay will result in echo and talker overlap – the problem of one talker stepping on the other talker's speech). Jitter is the variation in the inter-packet arrival time. As voice is sampled, it is separated into multiple packets. These packets may not arrive at their destination in the same sequence in which they were created, because some of the packets may have taken a longer route to reach their destination. Limiting jitter requires holding packets long enough to allow the slowest packets to arrive in time to play in the correct sequence. Packet loss occurs at times of peak loads and congestion, which is due to the link failure or inadequate bandwidth, whereby packets are dropped.

Research shows that users face difficulty in dealing with a delay in voice transmission of more than 150 to 200 milliseconds or a packet loss greater than 10%. There are numerous methods under development to improve voice quality in IP networks. Echo cancellers and jitter buffers promise to control delay and packet loss in IP voice transmissions. Engineering an IP network that minimizes end-to-end delay and packet loss is essential in providing acceptable quality of service.

Unfortunately, poor quality of service is a common occurrence in an IP Telephone System because of the dynamic traffic loads on the LAN. Efforts are underway to improve voice quality by developing features that use bandwidth allocation techniques to reserve the necessary bandwidth required for voice traffic. ESI's IVX is a prime example of a company who has solved this problem by installing a separate circuit board in each IP telephone that 'reserves' bandwidth to ensure voice quality. In addition, development is underway to improve the efficiency of standards used to packetize and compress voice. However, the wide-scale implementation of these features for most companies is still one to two years away. During this time, progress will be made so that reliable interoperability between different manufacturers' systems is assured. The bottom line is this; know what you are buying, its limitations, and current operating functionality within the scope of <u>your environment</u>. Then you will be able to set up your system to function, as you would have it.

Interoperability

Since standards for IP voice transmission have not yet matured, interoperability issues exist at all points in the IP network. Today, mainly proprietary voice compression algorithms are used to packetize voice. Standards such as Session Initiated Protocol (SIP) and H.323 are maturing. In the future a standard algorithm for packetized voice will solve the problem of interoperability between IP Telephone Systems, if adopted and adhered to. ESI's IVX has solved this problem with proprietary software that allows a simple 'open standards' protocol if you will, which allows them to operate with any current standard being used.

Capacity

Voice and Video both require a minimum bandwidth, including compression techniques to be useable. Therefore, network capacity must be calculated to cope with peak capacity under all circumstances.

Latency

As with capacity there are strict requirements for network latency. The voice and video packets must traverse the network in 50-100ms. Hardware latency as well as link latency must be calculated beforehand.

Quality of Service (QoS)

Different Applications; For example, if we treat voice and video as applications for the time being, then they require minimum standards as well. QoS standards such as 802.1q, 802.1p, DiffServ, MPLS must all be rigorously adhered if the goal is to be achieved. Recovery should be instantaneous. The planning should not only review predicted loads, but also review every element of the network. The purpose and failure analysis, mean time between failures, its effect on congestion, delay and utilization all has to be considered.

When planning the converged network, nothing should be viewed as trivial.

Also, the traditional enterprise PBX/Key System is renowned for its reliability. To date, IP Telephone Systems have not been able to approach this record for reliability. Before wide-scale implementation of IP Telephone Systems will occur, these systems must function as reliably as the traditional PBX. While recent advancements in PC and Network operating systems have helped close the reliability gap between the enterprise PBX/Key Systems and the LAN-based IP Telephone Systems, much of the reliability problems occur at the network hardware level. These failures can be attributed to poor network design, lack of bandwidth, routing mission critical voice traffic on commercial networks that are prone to lock ups, and poor interoperability between devices.

Services

There is one aspect that should not be underestimated. That is the effect of new applications and new IP telephony services. As we have witnessed with the explosion of the Internet on business, the use of IP telephony will generate more and more demand. After all no one owns the whole network, as each service provider can add services to its network, and each new application in the business will generate capacity issues. There is no finite end date associated with the move to a converged network, but the quicker the nettle is grasped, the greater the advantage.

Phone System Feature Limitations

Over the course of 30 years, the traditional enterprise PBX/Key System has evolved to support hundreds of different features. The large scope of available features makes the traditional PBX/Key, System highly customizable to meet the needs of a business. Today's IP-based phone systems provide customers with only a small fraction of the features that a traditional PBX/Key, System does. Today, businesses are purchasing IP systems for the sake of the technology, not for the depth and application of the communication features or business advantages offered. However, many companies have chosen to take their current systems – fully featured, and migrate that technology into their IP systems, which has given them a running start at the competition. ESI's IVX is one such example.

Regulatory Issues

Many states are now requiring business telephone systems to provide the exact location of a caller when a call to 911 is made. This is called E-911 Compliance. Currently, vendors (including Iwatsu America) have developed features to alert a local public safety answering point of the location of a caller making an emergency call. Circuit switched telephone systems can easily report the location by associating the origination point of a call with a specific system port.

There is no regulatory decree for 'Internet 911' at this time. In fact, if Internet 911 does become a state mandate, IP Telephone System vendors will have a unique challenge identifying the location of an emergency caller. When Simple Network Management Protocol (SNMP) is used to report the IP address of a device on the network, a map of devices can be generated. However, a recent trend has seen Dynamic Host Configuration Protocol (DHCP) used instead to assign IP addresses dynamically.

In a DHCP environment, the IP address associated with a specific device will change at random. Because of this, extra processing power will be required to create instantaneously accurate maps for the purpose of 911 reporting. This is a significant issue today that is not supported by IP Telephone Systems.

Development of a Converged Network

The convergence of voice and data systems and networks is rapidly becoming a reality. However, the wide-scale deployment of packet-based voice communications products will not begin to occur for three to five years. This time may be accelerated as interoperability standards evolve, and the Public Switched Telephony Network (PSTN) becomes more data-centric.

Eventually, the new packet-based telephony network will carry voice and data over a common transport, operating in parallel with the existing PSTN. New switches, routers, and access devices will populate the network allowing virtually seamless crossover from the PSTN to the packet-based telephony network. In this paradigm, the transport for telephony becomes transparent. Users will not be able to distinguish, nor will they care, whether a call is routed over a packet-switched or circuit-switched network. This new network paradigm will have the power to radically alter the way in which enterprise technology is deployed.

Innovative applications and business opportunities will be developed to take advantage of, and exploit the integrated delivery of voice and data over a single unified network fabric. Today, only a small subset of customers can justify a unique business need to adopt this new technology. As the network fabric and the technology mature, new applications and services will fuel the demand for IP Telephone Systems and other packet-based voice products.

Networking

The possibility of voice communications traveling over the Internet rather than the PSTN, first became a reality in February 1995 when Vocaltec, Inc. introduced its Internet Phone software. Designed to run on a 486/33-MHz (or higher) personal computer equipped with a sound card, speakers, microphone, and modem (see *Figure 7.1*), the software compresses the voice signal and translates it into IP packets for transmission over the Internet. However, this PC-to-PC Internet telephony works, only if both parties are using Internet Phone software.

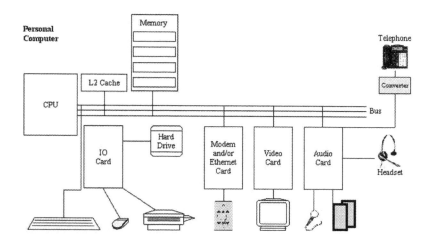

Fig: PC Configuration for VoIP

In the relatively short period of time since then, Internet telephony has advanced rapidly. Many software developers now offer PC telephony software but, more importantly, gateway servers are emerging to act as an interface between the Internet and the PSTN (see *Figure 7.2*). Equipped with voice-processing cards, these gateway servers enable users to communicate via standard telephones.

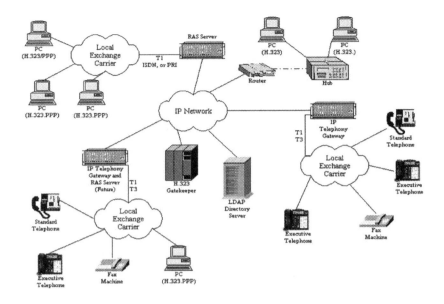

Fig 7.2 Topology of PC-to-Phone

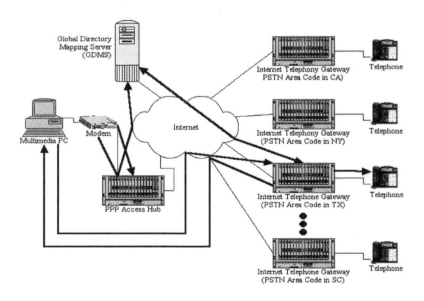

Fig 7.3 Sequence of VoIP Connection: PC-to-Phone

A call goes over the local PSTN network to the nearest gateway server, which digitizes the analog voice signal, compresses it into IP packets, and moves it onto the Internet for transport to a gateway at the receiving end (see *Figure 7.4*). With its support for computer-to-telephone calls, telephone-to-computer calls and telephone-to-telephone calls, Internet telephony represents a significant step toward the integration of voice and data networks.

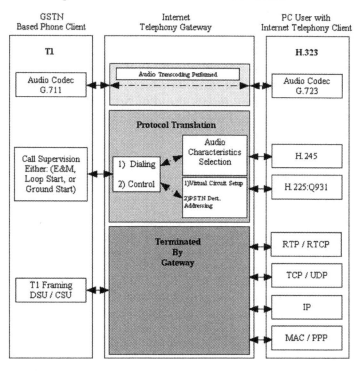

Fig 7.4: Sequence of VoIP Connection

Originally regarded as a novelty, Internet telephony is attracting more and more users because it offers tremendous cost savings relative to the PSTN. Users can bypass long-distance carriers and their per-minute usage rates and run their voice traffic over the Internet for a flat monthly Internet-access fee.

The ultimate objective of Internet telephony is of course, reliable, high-quality voice service, the kind that users expect from the PSTN. However, at the moment that level of reliability and sound quality is not available on the Internet, primarily because of bandwidth limitations that lead to packet loss. In voice communications, packet loss shows up in the form of gaps or periods of silence in the conversation, leading to a clipped-speech effect that is unsatisfactory for most users and unacceptable in business communications. However, should you wish to spend the money, you can eclipse this internet notion and go straight to point-to-point T1's, thereby guaranteeing your voice quality and virtually building your own in-house network. However, voice is a relatively predictable bandwidth user, verses data, which is subject to bursts. Therefore, you will still require a mechanism or control, which will guarantee, nay – safeguard your voice. ESI's IVX has such a built-in guardian, which will actually slow down the data in such an occurrence in order to preserve the quality of your voice transmission over VoIP.

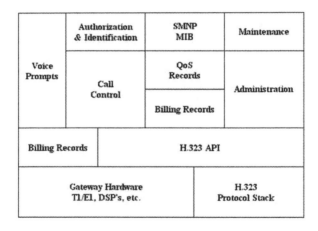

Fig 7.5: Internet Telephony

The Internet, a collection of more than 130,000 networks, is gaining in popularity as millions of new users sign on every month. The increasingly heavy use of the Internet's limited bandwidth often results in congestion, which in turn, can cause delays in packet transmission. Such network delays mean packets are lost or discarded.

In addition, because the Internet is a packet-switched or connectionless network, the individual packets of each voice signal travel over separate network paths for reassembly in the proper sequence at their ultimate destination. While this makes for a more efficient use of network resources than the circuit-switched PSTN, which routes a call over a single path, it also increases the chances for packet loss.

Network reliability and sound quality also are functions of the voice-encoding techniques and associated voice-processing functions of the gateway servers. To date, most developers of Internet-telephony software, as well as vendors of gateway servers, have been using a variety of speech-compression protocols. The use of various speech-coding algorithms – with their different bit rates and mechanisms for reconstructing voice packets and handling delays, produces varying levels of intelligibility and fidelity in sound transmitted over the Internet. The lack of standardized protocols also means that many Internet-telephony products do not interoperate with each other or with the PSTN.

Hardware, Software & Protocol Requirements

Hardware Requirements

- The exact hardware, which would be required, depends on organizational needs and budget. The list below highlights the most general hardware required.

- The most obvious requirement is the existence (or installation) of an IP based network within the branch office.

- A gateway is required to bridge the differences between the protocols used on an IP based network and the protocols used on the PSTN. The diagram below illustrates on a high level how the two types of networks are interconnected.

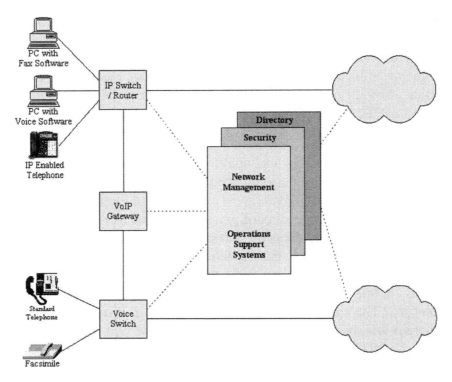

The gateway takes a standard telephone signal and digitizes it before compressing it using a Codec. The compressed data is put into IP packets and these packets are routed over the network to the intended destination.

- The PC's attached to the IP based network require the voice/ fax software outlined above. They also require Full Duplex Voice Cards, which allow both communicating parties to speak at the same time - as often happens in reality. These PC's also require the means to read in voice samples from users. Headsets or microphones are two devices used to take in samples. Similarly, the facility to hear the transmission must also be supported if microphones are used for input, then speakers can be used for output, however headsets provide both facilities and noise is reduced since headsets are private to the wearer.

- As an alternative to installing Voice Cards, IP Telephones can be attached to the network to facilitate Voice over IP.

- A secondary gateway should be considered as a backup in the event the primary gateway fails.

Software Requirements

- **Voice Processing Module:** This aspect of the software is required to prepare voice samples for transmission. The functionality provided by the voice processing module should support:

    o A PCM Interface is required to receive samples from the telephony interface (e.g. a voice card) and forward them to the Voice over IP software for further processing.

    o Echo Cancellation is required to reduce or eliminate the echo introduced as a result of the round trip exceeding 50 milliseconds.

    o Idle Noise Detection is required to suppress packet transmission on the network when there are no voice signals to be sent. This helps to reduce network traffic, as up to 60% of voice calls have silence and there is no point in sending silence.

    o A Tone Detector is required to discriminate between voices and fax signals by detecting DTMF (Dial Tone Multi-frequency) signals.

    o The Packet Voice Protocol is required to encapsulate compressed voice and fax data for transmission over the network.

    o A Voice Playback Module is required at the destination to buffer the incoming packets before they are sent to the Codec for decompression.

- **Call Signaling Module:** This is required to serve as a signaling gateway, which allows calls to be established over a packet switched network as opposed to a circuit switched network (PSTN for example).

- **Packet Processing Module:** This module is required to process the voice and signaling packets ready for transmission on the IP based network.

- **Network Management Protocol:** Allows for fault, accounting and configuration management to be performed.

Protocol Requirements

Although many aspects of IP Telephony remain un-standardized, some standards are now beginning to emerge. It is important when considering an IP Telephony solution, wherever possible the emergent standards are supported to facilitate interoperability with other existing products and future releases. There are many protocols in existence but the main ones are considered to be the following:

- **H.323** is an ITU (International Telecommunications Union) approved standard, which defines how audio/visual conferencing data is transmitted across a network. H.323 relies on the RTP (Real-Time Transport Protocol) and RTCP (Real Time Control Protocol) on top of UDP (User Datagram Protocol) to deliver audio streams across packet based networks.

- **G.723.1** defines how an audio signal with a bandwidth of 3.4 KHz should be encoded for transmission at data rates of 5.3 Kbps and 6.4 Kbps. G.723.1 requires a very low transmission rate and delivers near carrier class quality. The VoIP Forum as the baseline Codec for low bit rate IP Telephony has chosen this encoding technique.

- **G.711:** The ITU standardized PCM (Pulse Code Modulation) as G.711. This allows carrier class quality audio signals to be encoded for transmission at data rates of 56 Kbps or

64 Kbps. G.711 uses A-Law or Mu-Law for amplitude compression and is the baseline requirement for most ITU multimedia communications standards.

- **Real-Time Transport Protocol (RTP)** is the standard protocol for streaming applications developed within the IETF (Internet Engineering Task Force).

- **Resource Reservation Protocol (RSVP)** is the protocol, which supports the reservation of resources across an IP network. RSVP can be used to indicate the nature of the packet streams that a node is prepared to receive.

Integrating with Legacy Systems

IP telephony networks must be able to integrate with the legacy PSTN for seamless interoperability with the existing 160 million wire line telephones and 75 million wireless telephones. The IP equivalent SSP called a soft switch is being introduced to carrier networks. A soft switch is a highly distributed SSP based upon open systems standards.

As such, interoperability with the PSTN requires several soft switch interfaces to the PSTN. One of these soft switch interfaces is to the media gateway, which is a router that converts voice signals on the DS-0 trunk from the SSP into IP packets to the IP network. It compresses the 64K voice information into 16K blocks, which attaches an IP header with address information and sends the packets over the IP network. The traffic terminates at another media gateway, which performs the reverse process on the blocks. It converts the packets back into PSTN DS-0 equivalent voice messages. This all happens in the transport layer.

The Control Layer

PSTN and IP networks need to interoperate at the control layer as well. The Media Gateway Controller (MGC) is the call-control entity associated with IP Telephony and soft switches. The MGC and media gateway perform the job that the SSP does in the PSTN. The gatekeeper handles authentication, authorization, and address (AAA) mapping between the PSTN and IP network. On the PSTN side, people still are dialing PSTN telephone numbers. However, the IP networks routing are based upon IP addresses, so there must be an address conversion between the two spaces. The gatekeeper does that as well. A signaling gateway (SG) does the conversion between SS7-IP packets and the SS7 network. Interoperation between the two requires ISDN User Part (ISUP) trunk signaling information. When a call is made on the PSTN, the SSP (service switching point) sends an initial address message to the signal transfer point (STP). That message is communicated to the MGC via the SG. The MGC signals back to the SSP with an acknowledgement message, just as the SSP does today. In this way, companies can deploy networks based upon IP transport technology, but everything still will look like a normal SSP switch to the PSTN.

Also, the reality is that VoIP will be introduced into carrier networks when QoS, delay, application-specific standards and feature transparency issues are resolved. In addition, VoIP will have to meet the same challenge as VoDSL by integrating seamlessly with legacy switching equipment where required, as well as next generation equipment in the vast majority of cases. Through integration with emerging media gateways under soft switch control, VoDSL technology is evolving to meet the strategic needs of today's service provider and ensuring that heterogeneous network solutions truly work together.

Chapter Eight
IP Telephony Services Market

---

a) Market Overview

b) Market Segments

    i)   IXC's (InterExchange Carriers)

    ii) CAP's (Competitive Access Providers)

    iii) RBOC's (Regional Bell Operating Companies)

    iv) CLEC's (Competitive Local Exchange Carriers)

    v) ISP's (Internet Service Providers)

    vi) ITSP's (Internet Telephony Service Providers)

c) Emerging Trends & Technologies

d) Challenges & Constraints to IP Telephony

e) Future Growth

IP Telephony Services Market

Market Overview

The telephony market is currently dwelling in the initial stages of a paradigm shift that is transforming the delivery of voice services throughout the world. The Public Switched Telephone Network (PSTN) has been the most reliable medium for connecting phone calls for over 80 years now. However, the recent technological advancements and deregulation of the telecommunications industry has cost a lot to this copper-wire legacy system. The specially designed dedicated circuits to carry voice traffic are limited.

Internet Protocol (IP) has made a similar impact on the telecommunications market as in the data world. With IP-based voice transmission, the medium is immaterial. In simpler terms, voice is now data, which can be transmitted over any network, including cable, fiber, wireless and copper. The future of voice technology is centered on IP; its inherent reliability and flexibility.

IP-based voice products have shown a rapid rise in gaining the market share by catering to such problems with a barrage of exciting new applications and capabilities that will continue to proliferate. These additional features and functionalities work in a manner to stabilize carrier profitability, both in terms of operational savings and new revenue sources. Carriers are able to evolve their offerings from mass-market to a more specialized tailor-made offering, thereby addressing the specific needs of user-types or vertical segments. This has a significant impact on service differentiation and customer retention

Voice over IP (VoIP) has been progressing ever since it was first proposed in the mid 90's. Many businesses have hence adopted it for voice trunking and more so for desktop voice.

The main reasons behind this are:

- Lower ownership cost in totality: An integrated voice and data system saves an enterprise from having to deploy and maintain two distinct services.

- Application support: It is now possible to create and put in to use simple but highly effective IP phone-based applications integrated with database and simple instructional data.

With a few exclusions, to date there has been far less interest and success in the consumer VoIP market. Many service providers are offering a range of completely distinct offerings, both 'on net' (PC-to-PC) and off-net (PC-to-phone, and increasingly phone-to-phone). However, overall traffic and hence revenues generated are quite low as compared to the overall voice market.

Nevertheless the potential of VoIP is lucid enough, and it still attracts service providers in taking ever-greater proportions of both long-distance and particularly, international traffic. Significant market development still faces a lot of barriers. This briefly summarizes their impact on the scenarios and their influence on the strategies of the key players in VoIP services.

However, without much difference in service offerings, prices have continued to fall. Such problems are largely due to the carriers' use of outdated circuit-switched technology. Also that the proprietary nature of such technology does not allow for innovation and better service management.

With an already established Internet network, which is important for VoIP communications, it is in a perfect place and getting better everyday. However, the need for VoIP handsets and its accessories has emerged only lately. At this point in time, organizations like Cisco Systems, Siemens, Northern, ESI and Avaya and Northern, are heavily investing in VoIP with the goal of becoming the dominant player/leader of this fast growing/developing industry. However, many companies such as Avaya have held a leadership position in analog PSTN based enterprise telephony, which should provide them a strong corporate network that could help them to be successful, but sometimes – and in fact to date, have limited their innovation.

Market Snapshot

- The consumer VoIP market is presently a small portion of the overall voice market, comprising of a few localized large operators with fewer other smaller players.

- Service providers are currently focusing on key destinations where slow de-regulation has kept charges high, and new markets where the penetration of telephony is low.

- The cable industry is keenly eyeing this opportunity of VoIP.

- Various issues need to be dealt with before the consumer VoIP market can achieve any major breakthrough. The key factors being the quality of service, and some uncertainty over regulation and marketing.

- However, most incumbent providers have ignored and resisted the natural transition into VoIP technology, which means that they are in for a long-term threat in terms of revenues.

Market Segments

There is a wide range of numbers, which illustrates the present dimension of the IP telephony market and its growth over the next three to five years. While specific projections vary, even the most conservative analysts are predicting phenomenal growth.

The summaries of the numbers are as follows:

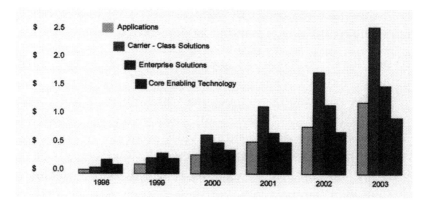

Source: Piper-Jaffray, IP Telephony, Driving the Open
Communications Revolution

Traditional telecommunications is going through a transitional phase. High fixed costs, slow revenue growth, and the lack of capital for capacity expansion are the main reasons why many large carriers are outsourcing an increasing volume of international traffic to wholesale carriers. Unfortunately, the same economic factors, coupled by the slimmer margins of the wholesale model, have compelled many traditional wholesale carriers to just leave the market altogether. With the surfacing of high quality global VoIP networks and strong VoIP solutions for the enterprise, Internet telephony is proving to be a system par excellence to the more expensive, closed, legacy networks.

Within a decade, VoIP has evolved from a gimmick that only a PC hobbyist could love, to a major force in international telecommunications. IDC data reported that in 2001 more than 10 billion minutes, or 6% of international traffic, was transported over IP networks and predicted that international VoIP minutes will reach 78 billion minutes in by the end of the year 2005. Based on current run rates, two of the ten largest carriers of international voice traffic today are VoIP carriers - iBasis and ITXC. Moreover, it is not a matter of small calling card operators and Tier three carriers trading voice quality for the cost savings of VoIP. More than 70% of iBasis' traffic in the second quarter of 2002 came from Tier One carriers, which is the world's largest and most demanding.

The VoIP market segment is expanding as more and more carriers, incumbents and new carriers, are transitioning TDM networks to IP networks, making interconnections with wholesale VoIP carriers easier to implement, both politically and technically. In addition, the cost of building or expanding VoIP networks is dramatically lower than that of a circuit switched network. VoIP carriers can add new routes and capacity on existing routes by adding lower cost equipment and IP connectivity. VoIP network deployment costs come in increments of $50,000 to $100,000, rather than the '$1 million plus' increments of telco switching facilities and dedicated circuits. Developing countries find VoIP economics very compelling and are embracing the technology to expand their domestic communications infrastructure at perhaps 10% of the cost.

From a technology perspective, VoIP equipment has improved performance, capacity, and service quality dramatically over the past five years. Yet, to provide the high quality voice service and reliable call completion required by Tier One carriers, still requires much more than off-the-shelf components and Internet access. Proprietary patent-pending software technology is necessary to turn even the most state-of-the-art VoIP technology into a successful production network.

VoIP carriers are constantly implementing improvements in routing, traffic monitoring, management, in network design and equipment, all to enable the deployment of more efficient and reliable VoIP infrastructures. It is truly remarkable to consider how fast VoIP has emerged. Just a little over five years ago, international VoIP carriers were little more than an idea and a PowerPoint presentation. Today, networks like iBasis span more than 650 points of presence and 85 countries. Indeed, iBasis is ranked as the number one international wholesale carrier, among carriers of any kind, according to the Atlantic-ACM International Wholesale Report Card 2002. As fast as the VoIP segment has grown, it has still only begun to penetrate the global telecommunications market. With growing enthusiasm for VoIP at the enterprise level and the increasing use of VoIP for market-proven enhanced services like conferencing, the next five years promise to be just as remarkable.

IXC's (InterExchange Carriers)

IXC's (InterExchange Carriers), both within the US and internationally popular ones, include AT&T, MCI, Sprint, Verizon, Deutsche Telekom, Telecom Finland, Telecom New Zealand, and Daewoo International. AT&T's Globalnet division is currently offering IP telephony service, MCI has already begun building PC based web servers, which may support IP telephony, and Sprint has announced its Global One service that offers dedicated bandwidth on demand.

Even telephone carriers not directly involved in the Internet business will most likely offer services that take advantage of IP telephony, such as reduced-rate voice and fax, and store and-forward fax and voice messaging.

CAP's (Competitive Access Providers)

There are a multitude of small national and international carriers who offer an alternative to the major carriers for long distance minutes. These CAPS now comprise over 3/10th of the international long distance market. Today, the strongest play for IP telephony in the CAP market is in international long distance where arbitrages between the US and nationalized carriers are artificially high. Both voice and fax are major requirements.

RBOC's (Regional Bell Operating Companies)

While the ruling allowing RBOC's, such as; Bell Atlantic, SBC and US West to offer long distance services are still in dispute, it is natural to assume that RBOC's will look at packet-based technology as they begin to roll out their inevitable long distance applications. Additionally, through their alliances with ISPs, RBOC's are a natural convergence point for traditional voice communications and data networks.

CLEC's (Competitive Local Exchange Carriers)

CLEC's have a strong reason to adopt VoIP technologies into their networks. Most CLEC's are focusing on specific client types such as commercial businesses instead of home subscribers. As such, CLEC's are growing rapidly in metropolitan regions while not maintaining any network infrastructure between regions. It is common for a CLEC to have service in Los Angeles and San Francisco but must use AT&T or other IXC's to move voice between the CLEC regions.

With IP telephony, the CLEC can easily connect their regions together over a private IP network allowing low cost intra-Network calling as well as fax services. IP telephony gateways can become the glue to merge their metro PSTN networks together instead of relying on other PSTN service providers who charge an exorbitant rate for access. CLEC's are a fast growing market with over 2,500 applications for CLEC status with the FCC.

ISP's (Internet Service Providers)

There are arguably between one and two dozen large ISP's, notably Alternet, BBN Planet, Digital Express, NetCom, PSI, and MCI WorldCom. There are also hundreds to thousands of smaller and single region ISP's such as PANIX and TIAC, as well as tens of thousands of bulletin board services (BBS's). Many international ISP's are also offering IP telephony services, such as Rimnet of Japan and OzE-mail of Australia. Many ISP's are moving towards CLEC status.

ITSP's (Internet Telephony Service Providers)

Representing a new class of service providers, these companies are building global IP networks specifically designed for low-latency traffic, including voice and fax. This emerging market already had established leaders including ITXC, Gric, Delta Three, and Telic. net, and will undoubtedly spawn new competitors in the coming months.

Emerging Trends & Technologies

Internet Protocol (IP) telephony, also known as Voice over IP (VoIP), is becoming a key driver in the evolution of voice communications. VoIP technology is useful not only for phones but also as a broad application platform enabling voice interactions on devices such as PC's, mobile hand held, and many vertical-specific application devices where voice communication is an important feature.

Several factors will influence future developments in VoIP products and services. Currently, the most promising areas for VoIP are corporate intranets and commercial extranets. Their IP–based infrastructures enable operators to control who can, and who cannot use the network.

Another influential element in the ongoing Internet-telephony evolution is the VoIP gateway. As these gateways evolve from PC– based platforms to robust embedded systems, each will be able to handle hundreds of simultaneous calls. Consequently, corporations will deploy large numbers of them in an effort to reduce the expenses associated with high-volume voice, fax, and videoconferencing traffic. The economics of placing all traffic, data, voice, and video over an IP–based network will pull companies in this direction simply because IP will act as a unifying agent, regardless of the underlying architecture (i.e., leased lines, frame relay, or ATM) of an organization's network.

Commercial extranets, based on conservatively engineered IP networks, will deliver VoIP and facsimile over Internet protocol (FAXoIP) services to the general public. By guaranteeing specific parameters, such as packet delay, packet jitter, and service interoperability, these extranets will ensure reliable network support for such applications.

VoIP products and services transported via the public Internet will be niche markets that can tolerate the varying performance levels of that transport medium. Telecommunications carriers most likely will rely on the public Internet to provide telephone service between geographic locations that today are high-tariff areas. It is unlikely that the public Internet's performance characteristics will improve sufficiently within the next two years to stimulate significant growth in VoIP for that medium.

Cost savings for both customers and carriers are driving the near-term growth of the VoIP market. As packet-switched voice displaces the older circuit-switched voice technology, VoIP carrier revenues will explode, fueled by lower operational costs to carry a voice call along with new service offerings such as follow-me, unified messaging, and multimedia communications.

According to the prevailing market research, the share of international traffic that is migrating to VoIP networks is approximately doubling each year. While international minutes of use are projected to grow at roughly 15% annually, international VoIP minutes of use are projected to grow at a much faster rate. iBasis and ITXC both claim to carry a double-digit share of the traffic on many prominent international routes including US to Mexico, China and Russia. VoIP has also continued to grow in developing countries, where the technology experienced its initial success against the PSTN networks that were unreliable and expensive to expand.

However, the public Internet will be able to handle voice and video services quite reliably within the next three to five years, once two critical changes take place:

- An increase by several orders of magnitude in backbone bandwidth and access speeds, stemming from the deployment of IP/ATM/synchronous optical network (SONET) and ISDN, cable modems, and x digital subscriber line (xDSL) technologies, respectively

- The tiering of the public Internet, in which users will be required to pay for the specific service levels they require

- On the other hand, FAXoIP products and services via the public Internet will become economically viable more quickly than voice and video, primarily because the technical roadblocks are less challenging. Within two years, corporations will take their fax traffic off the PSTN and move it quickly to the public Internet and corporate Intranet, first through FAXoIP gateways and then via IP–capable fax machines. Standards for IP–based fax transmission will be in place by the end of this year.

Throughout the remainder of this decade, videoconferencing (H.323) with data collaboration (T.120) will become the normal method of corporate communications, as network performance and interoperability increase and business organizations appreciate the economics of telecommuting. Soon, the video camera will be a standard piece of computer hardware, for full-featured multimedia systems, as well as for the less-than-$500 network-computer appliances now starting to appear in the market. The latter in particular should stimulate the residential demand and bring VoIP services to the mass market by including the roughly 60 percent of American households that still do not have a PC.

Challenges & Constraints to IP Telephony

Unlike traditional public switched telephone networks (PSTN's) and cellular networks, the Internet originally was not designed as a dedicated real-time network for voice communications. On the contrary, the Internet was designed as an asynchronous data communication network, allowing data packet loss and retransmission, without dedicated bandwidth for each user. Furthermore, a single and centralized operator that can coordinate the flow and quality of caller interactions does not manage the Internet. Unlike PSTN's and cellular networks, the Internet consists of disparate networks and service providers, which adds to the difficulties of providing real-time communication services. This combination of factors makes the Internet a challenging network medium for real-time communication scenarios such as voice conversations.

Implementing VoIP takes IT managers into a new realm and consequently requires a different focus of attention:

- Users expect high voice quality and reliability. Telephones have been around a long time and the technology has been perfected to demanding expectations. Even digital cell phone users expect high quality. Hence, users expectations are high and even the slightest drop in quality and reliability of VoIP will be noticed and complained about.

- Voice gobbles up bandwidth. Managers need to allocate adequate bandwidth to accommodate voice. Ultimately, inadequate bandwidth reduces network performance and degrades both data and voice transmission. However, compression and priority routing have proven to be useful strategies.

- Talent in monitoring and managing data networks doesn't necessarily transfer to VoIP. Training in-house managers or contracting with outside professionals should be considered. There will be surprises.

- A unified infrastructure lowers the costs of communication for end users and service providers. Using network management software to simultaneously monitor performance metrics of voice and data enhances the performances of both. It's usually cost prohibitive to run two separate networks.

VoIP market growth largely depends on three factors:

1. Increasing consumer awareness of quality and cost savings.

2. Constantly improving cost-performance of underlying technology.

3. Transparency in standards and protocols among equipment providers

One of the greatest challenges for IP telephony is to develop networks, which are not only scaleable but also seamless to the subscriber and to the service provider. If the service is difficult to access by the subscriber due to complex dialing pLAN's and special PIN numbers, or requires significant time to complete a call, or has constant call drops, then the IP gateway will only be used by a limited client base.

If it is difficult for the service provider to install, administer, settle, and bill, it will have longer ROI times and will be less likely to be deployed. Incumbent carriers have very specialized billing systems already in place and are not likely to create an entirely new billing system just for IP telephony. An SS7-based solution allows legacy OSS and OAM&P systems to be utilized saving considerable expense. What is required is IP telephony switches with SS7 call control features that can be deployed inside the PSTN. This allows seamless subscriber interaction such as one-stage dialing and instant call setup. Traditional carriers with legacy billing and settlements can use it, since it can generate SSP-style CDR records. Next-generation carriers who wish to use an IP telephony billing service can also use the IP telephony switches.

Future Growth

The Telecommunications Industry Association (TIA) predicts spending in the voice processing equipment and services market will rise at a healthy high-single-digit growth rate in 2003 – following two years of flat growth – and then at double-digit rates through 2006, according to its recently released 2003 Telecommunications Market Review and Forecast.

The weak economic environment and cutbacks in corporate investment were primary causes of the slowdown in spending over the past two years. However, continued growth in the call center market, a rebound in PBX system shipments, gradually improving economic conditions and the need to replace aging equipment with systems that incorporate new technologies will drive growth. Voice processing market spending is expected to increase 9.7% to $7.6 billion in 2003 and to reach $11.2 billion by 2006.

Growth Forecast on the basis of Market Segmentation

- **Every voice processing market segment** will feature steady improvement in investment over the next few years, including voice mail equipment. Commonly tied to the PBX market, the voice mail segment experienced a 26 percent drop in sales between 1999 and 2002, after PBX system purchases were accelerated in association with Y2K. The surge in popularity of IP PBX's will lead to a jump in PBX system sales, which will propel the voice mail equipment market. Voice mail equipment spending is expected to total $2.0 billion in 2003, up 5.9% from 2002.

- **The interactive voice response (IVR) equipment market** is also expected to receive a boost. New applications that make IVR user-friendlier should lead to 4.8 percent growth in 2003, reaching $1.3 billion in revenue. Voice recognition technology is the fastest-growing component of the IVR market and will further stimulate growth.

- Spending on **automatic call distributors (ACD's)** rose by 6% in 2002 to $1.7 billion, thanks in part to continued growth in the call center market. Increased spending on new skills-based products will keep spending up despite its having achieved market maturity. The ACD market is expected to average 5.9% growth on a compound annual basis through 2006, reaching $2.1 billion.

- **Unified messaging software and services** will also contribute to overall industry growth. As technologies are developed to convert text to voice more efficiently and voice-to-text applications emerge, demand for unified messaging should soar; historically, technologies that save time have been very successful. By 2006, spending on unified messaging is expected to rise to $4 billion from $1.3 billion in 2002. Growth will average 32.4% compounded annually.

- Automatic call distributors, predictive dialing and unified messaging are three of the strongest categories of spending in the voice processing equipment and services market.

- Spending in the **voice processing equipment and services** market is expected to increase at a 12.9% compound annual growth rate through 2006. In 2003, voice processing equipment spending is expected to total nearly $6 billion – up 6.6% from 2002, while unified messaging software and services is expected to increase 23.1% from 2002 to $1.6 billion, for a total market size of about $7.6 billion.

  Predictive dialing, the smallest segment of the voice processing equipment and services market, will achieve 12.8% growth in 2003 reaching $880 million, up from 8.8% in 2002. Predictive dialers serve the outbound telemarketing industry, the component of call center activity most closely tied to revenue generation. Consequently, in a tight economic environment, demand will remain strong through 2006.

- Associated expenditures on the **operations support systems (OSS's)** used to create and maintain VoIP-based services will also increase over the forecast period at a compounded growth rate of 58%, indicating heavy investment in correlation with the phenomenal revenue growth of VoIP.

- The revenues from the **wholesale calling card services market** between 1997 and 2004 are defined as the revenues that are received through the sale of wholesale calling card services to the resellers. Revenues for 2000 were approximately $512 million, representing less than 1 percent growth from 1999. This is expected to decrease as price declines for using packet networks will outweigh minute growth throughout the period, mainly due to price pressure from Tier 1 providers offering their own prepaid cards at just above cost and without a high premium. The expected compound average growth rate (CAGR) for wholesale calling card revenues is expected to be negative into the double digits. Potentially, calling card market revenues are in slight decline today, as a number of prepaid providers have shifted an increasing number of voice volumes to packet networks. By 2004, it is expected that a majority of wholesale calling card minutes will be transported over packet networks, significantly reducing potential revenue through price erosion. As a result, expect yearly double-digit declines for wholesale calling card revenues beyond 2004. Even though the carriers are going to be hurting a bit, this is good news for resellers because the majority of lost revenue will come from lower prices.

• **Softswitch** is a key element in the growth of IP (Internet Protocol) networks and services, including voice over IP. Financial analyst firm Probe expects the market to grow at an over 50% CAGR from 2002 to 2008, when it will be a $5+ billion market worldwide. Among diversified vendors, Siemens, Nortel and Ericsson are the current leaders. But, according to Christine Hartman, research director for PFA, one or two large deployments can still significantly change the picture. Among newer players, Hartman suggests that UT Starcom, Italtel, Sonus and Vocaltec are significant providers. In addition to strong market growth, PFA expects above average M&A activity in the softswitch area in the next 3 years.

Insight's newly-released report on IP Telephony: Service Revenue and OSS Expenditures for Voice over IP Networks 2002-2007, reveals that VoIP services will grow at a compounded rate of over 72 percent over the forecast period, making packet voice services one of the fastest-growing segments in the battered telecommunications segment. From a mere $13 billion in 2002, voice over packet (VoIP)-based services will grow to nearly $197 billion by 2007, according to the study.

The overall United States telecommunications industry, including equipment and services, generated more than $600 billion in revenue in 2000. While VoIP is currently a small fraction of this, it is growing quickly. In North America, wholesale VoIP sales were estimated to approach well over $400 million in 2002. Total equipment purchases of VoIP gateways, soft switches such as IP Private Branch Exchange (IP PBX), and VoIP application servers are expected to reach almost $12 billion by 2006, a six-fold increase over 2001. Similarly, the revenue from selling wired enterprise IP-phones may be in excess of $2.7 billion by 2006 (this figure does not include mobile IP phones or phones used in private homes).

Chapter Nine
Putting Your Entire Project Together

---

a) IP Telephone and our Customers

b) The company's I use and trust to make my IP network bullet proof

i)   ESI (Estech Systems, Inc.)

ii) Maher Technology Group

iii) Itelisys

iv) FaxBack

v)   IdentaFone

vi) Tapit

vii) Sureen Systems

viii)Your IP system at home

What is IP Telephony? ~ *and* ~ Who Are Our Customers?

IP Definition:

Internet telephony (IPT) is the transport of telephone calls over the Internet. It does not matter whether they are traditional telephony devices, multimedia PC's or dedicated terminals, which take part in these calls.

Telephony technology, which can be enabled across data networks through Internet protocol, is known as IP Telephony. This is characterized by the set of standards which enables voice, data and video to coexist over IP based LAN's, WAN's, Frame Relay networks, ATM backbones and the Internet. The applications of which include PC-to-PC communications, PC-to-phone and phone-to-phone communications. The convergence of the circuit switched networks such as the Public Switched Telephone Network, or PSTN with packet switched networks such as the Internet, intranet, LAN, WAN and other network types, best explains the nature of IP Telephony, which has created new applications and revenue opportunities thus enhancing the existing communication of systems.

IP Customer Definition:

An IP customer quite simply is anyone who has more than one location, work environment, or multi-site communication requirement.

This may sound relatively simple or even superfluous at first glance, but not if you look closely at the current environment in which we find ourselves today, both in the way that business is evolving and the explosion in low-priced data, voice and data, and DSL services.

Let's delve further into some of the possibilities that you may run across and yet not clearly see the simplicity of the IP solution. Yes, the pure IP answer for a company is to link two, or several locations together. However, if you wait for this scenario to come along, with the associated and available funds from the company's coffers, while trying to justify the sale by stating, *"It's free long distance between the two sites*", you will surely starve to death. The true IP solutions of today are created and are far more underlying than the more obvious solutions that you are currently attuned to.

So, starting out easily, let's not overlook one location with a home office. What about one location and ten ACD (Automatic Call Distribution) employees all working from their homes? Have you ever come across a SOHO with five partners whom also work in SOHO environments? One that I did recently was a customer that had no other offices, no home offices and no need of any kind whatsoever for IP. This continued right up until we began discussing his customers and how reliant he was upon certain accounts for his business to be profitable. Competition, as with any industry is fierce, his buyers were not local and they were in constant contact with his various departments. The solution? He located an IP telephone with all of his main buyers so they could be part of his business system. They could intercom back and forth, received voice mails in real time, one touch to any of his departments, no long distance charges, pure convenience, etc. He demonstrated how important their business was to him, and in return, they picked up his phone more often. After all, his telephone was already on their desk.

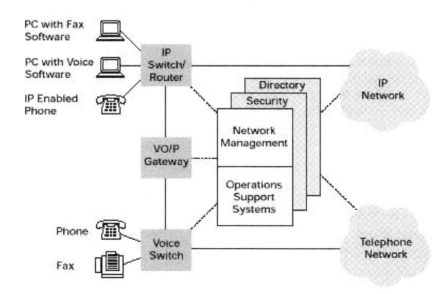

With tenure of only five years, the entire market of IP telephony has evolved and has become a lucrative business worth billions, the maxim of which is to provide service to carriers and millions of individual consumers with millions of individual needs. You need to create them!

Included is a simple five (5) location IP solution for a 2 location site with 3 remote SOHO's.

Still there can be other daunting jobs to figure out. One such company had five sites to link together. They wanted full integration between sites, facsimile distribution between all of their locations – at the desktop level, utilize their current Microsoft Outlook with unified messaging and screen synchronization, full call accounting with polling capabilities from each site and a completely seamless operation with the convergence of both voice and data. Additionally, they wanted their own in house software customized to their particular operations. Here is what we offered them.

- ♦ ESI – Five IVX Telecommunications systems with VoIP capabilities – 24 channels per location

- ♦ Maher Technology Group – Full integration of all T1 voice circuits, T1 facsimile circuits and T1 data circuits throughout their entire network.

- ♦ Intelisys – Four point-to-point T1's between locations to the head office, one T1 for their facsimile network and one bonded T1 at the head office to support all of their Internet traffic.

- ♦ FaxBack – A system that will run over their current environment / Microsoft Outlook, utilizing the Internet to virtually allow any of their users to send a fax from any computer station / laptop as long as they had an Internet connection, utilizing DID's through the T1.

- ♦ Identafone – 226 screen synchronization license's for their user's to splash an inbound caller onto their computer screen and then pop-up a Microsoft Outlook contact from their database.

♦ Tapit – a call accounting program at each location, utilizing polling in order to maintain a complete call accounting structure of billing per person, department, branch and organization.

♦ Sureen Systems – Full integration of all desktop, LAN, WAN systems to communicate as one system.

The whole system put together looked something like this.

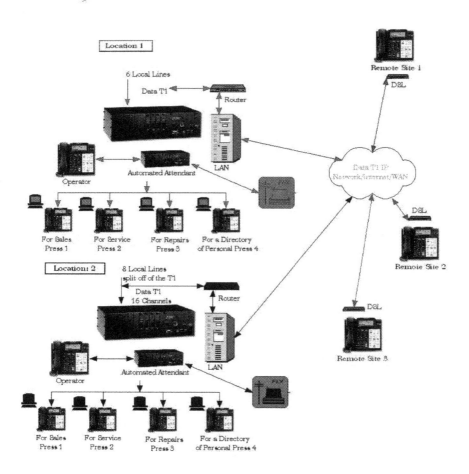

DataNet) **Possible Five Site Configuration Using IP**

The company's I use and trust to make my IP network bullet proof

Esi / Ivx / Estech Systems, Inc.

ESI has been around for a long time and started out as an OEM for other company's products. I first encountered them in October of 1996, when their first telecommunications product under their own name had just been released. It was called the IVX 1634 (now referred to as 'the classic'), which was a digital cabinet with 16 lines, 32 analog telephones, 2 analog ports, 8 ports of voice mail and 17 hours of storage. For its time it was a great little system, which provided an 'all-in-one solution' for small business's.

Since that time, ESI has come a long way with many product releases, from the 128, to the 128*e*, 72*e*, IP 200, IP 40, the 'S' Class, IP 40*e*, IP 200*e* and most recently the IVX 'X' class, which allows for 252 call processing ports in various combinations, supporting up to 168 telephones and up to 3 T1 circuits. Their voice mail has also grown to 24 ports and 420 hours! What is truly unique about the IVX product line is still their dedication to the 'all-in-one solution' for customers. With most product lines, if you want an ACD application – well, that's extra. More than 4 ports of voice mail – you'd better get out your checkbook. IVX has managed to continually add and grow their product line, while maintaining the approach of keeping the customer continually satisfied by giving them access to all of the features of the system – and all for one price.

I have sold a great many systems in my time, and here is where ESI really shines. It is a solid fact, that products do fail and that software does have 'bugs'. It is completely unavoidable. First off, the IVX is one of the only manufacturers that offers a five (5) year warranty directly from the manufacturer. How safe is that! Your vendor goes out of business and you can still go to the manufacturer to find another company to do your service. If you don't like your service company, you can get someone else to do your service and your warranty is still in tact. That is phenomenal!

As well, with most systems, if there is a 'bug' in the software, you will get a letter in the mail stating that they recognize the problem and there will be a fix in the 3rd quarter, which does you very little good if you are in the 1st quarter, and this was a feature that your company urgently required. I have personally discovered a dozen bugs in various different IVX systems. Most of them were fixed within a couple of days to a week, while just one took over 2 weeks. Their engineers are in house and at their disposal. This is key to their customer service and huge in terms of customer satisfaction.

So, after saying all that and wondering if that wasn't enough (it isn't) the IVX 72*e*, 128*e*, IP 200*e*, and the 'X' class systems are all IP ready and IP capable! IVX has taken the guesswork out of installing IP systems and transitioned it into an art. Your options in this environment span from a true IP end-to-end solution, or an IP based system, which utilizes a 3, 12 or 24 port voice path between systems, or a gateway from system to system, which will link non-ESI systems together with ESI/IVX cabinets. IVX can actually link up to 100 systems together! If you are a national company, just imagine the savings.

What's more, their programming is the most simplistic I have ever had the pleasure of using. One piece of software programs both the voice mail and the telephone system at the same time in an easy to use windows format. And, if your lines are down, you can still program the system via an Internet connection!

The IVX system will satisfy most company's concerns with respect to functionality, ease of use (which is outstanding), feature rich environment, warranty, security, and a price point that would please even the most frugal buyers. Below is some information directly from the manufacturer.

ESI (Estech Systems, Inc.) designs, manufactures and markets advanced, highly featured **business communications systems**. Progressive vision, leading-edge technology, legendary reliability and award-winning products define ESI. Since its inception in 1987, ESI — a privately held corporation based in Plano, Texas — has shipped over 100,000 systems, and has enjoyed remarkable growth and uninterrupted profitability.

ESI is proud to be an **ISO 9001:2000-certified** company. Beyond mere certification, the stringent ISO quality assurance standards are inherent to ESI's business philosophy and everyday practices.

ESI is the first manufacturer to build a truly **combined** telephone and voice mail system, and has numerous patents granted or pending on its products' unique design and features. ESI is highly regarded for supporting unique, real-world applications with practical and effective features. **How** does ESI accomplish these feats? By using **digital signal processors (DSPs)** — high-powered semiconductors that manage analog functions in a digital environment — and **ESI's innovative system software** to replace hundreds of hardware components. This enhances product reliability. It also makes the design and production of new systems, as well as the upgrading of existing systems, more efficient and cost-effective.

A brief history: ESI's advanced communication systems for business

In 1995, after several years of success in producing voice mail products, ESI used its advanced voice storage and DSP software development skills, combined with phone system hardware design expertise, to create the IVX® All-In-One Digital Phone System — the first telephone/voice mail system based on a single processor. Its great success led in 1999 to the introduction of the larger and even more fully featured IVX 128 system.

ESI became an early leader in the exploding **VoIP** (voice over Internet Protocol) marketplace in 2001 when, after years of extensive research and testing, it began shipping its **IP Series All-In-One IP Phone Systems**, as well as an IVX Series model that could be upgraded to use VoIP. Also in 2001, ESI released the **Remote IP Feature Phone**, followed in 2002 by **Esi-Link** — offering multi-location customers the convenience and cost savings of networked communications. These VoIP solutions are unique in their simplicity of support and administration.

In 2003, ESI began shipping an all-new generation of business phone systems — **E-Class** and **S-Class** —and, in 2004; these highly successful products were further enhanced with the **Generation II** systems, which have even greater application flexibility. The new, larger **X-Class** system also was announced in 2004, along with *VIP* — a call management solution which lets users manage their voice mail, contacts, and programming in conjunction with *Microsoft® Outlook®*.

Patented higher-performance technology

ESI has many patents either granted or pending on its systems' design advances. One patent covers the single-processor integration of telephone and voice mail systems. Another protects ESI's ability to keep Caller ID data with a voice message as it moves within the system, which allows easy callback, easy speed-dial entries and many other highly convenient features. (See *http://www.esi-estech. com/news.*)

ESI's innovative products use significantly less hardware, resulting in **increased reliability**. As a result of all these features (and many more), ESI products provide **significant benefits** and **greater long-term customer satisfaction**.

ESI is a leader in **secure** IP-based telephony solutions for companies of all sizes. ESI's VoIP design relies on the company's **custom-developed operating system**, and allows **only safe communications** to reach the system.

Products

ESI phone systems offer special ESI innovated-features such as live call recording, live call screening and the Verbal User Guide™. Beyond those similarities lie some differences:

- The **X-Class and IVX E-Class Generation II systems** are ESI's flagship products. Each offers a selection among multiple **Digital Feature Phones**, cost-effective expansion capabilities, automated attendant, automated call distributor (ACD) and voice mail. Also, these systems boast extensive expandability, with up to 252 call-processing ports and up to 420 hours of voice storage, along with support for T1, ISDN PRI, and TAPI computer/telephony integration (CTI). IP capabilities can be added to IVX X-Class and IVX E-Class, allowing existing customers to benefit from VoIP technology.

  - **IP E-Class** includes **IP 200e** and **IP 40e**. These are advanced IP network-based business telephone systems, combining the rock-solid performance and acclaimed telephony features of IVX E-Class, and the state-of-the-art delivery of packetized voice to the desktop via the existing LAN/WAN.

- ESI's **Remote IP Feature Phone** connects via broadband back to an IP-enabled ESI phone system at your main office. It provides the same functions as an ESI Feature Phone actually located at the office. This makes it ideal for teleworkers, small satellite offices and executives working from home.

- **Esi-Link** allows up to **100** IP-enabled ESI phone systems to be connected across an enterprise's WAN while sharing the advanced ESI feature set. The optional ESI **IP Gateway** allows connection of non-ESI phone systems to the customer's Esi-Link network.

- **IVX S-Class Generation II shares many fine features with IVX E-Class Generation II** (refer to *www.esi-estech. com/chart*) but is designed for simpler applications. IVX S-Class Generation II:

  - **Grows** to 16 CO lines, 32 digital stations and eight analog stations.

- **Uses the same outstanding Digital Feature Phones** as IVX X-Class and IVX E-Class. This makes it easier and less expensive to upgrade when your company's communications needs change.

- Includes **voice messaging flexibility** to meet customers' unique needs: either the same full-featured voice mail/auto attendant as X-Class/E-Class models or, for simpler business environments, ESI's unique Integrated Answering Machine.™

- *VIP* software — an option for X-Class, E-Class Generation II, and S-Class Generation II systems — is a call management solution, which lets users manage their voice mail, contacts, and phone programming in conjunction with *Microsoft Outlook*.

Why consider ESI?

- Customer-focused product development — ESI consistently incorporates the newest advanced and viable technologies into usable, real-world products — and enhances those products based on what you, the customer, says.

- **Commitment to total customer satisfaction** — ESI understands the needs of real-world businesses like yours, and designs systems with you in mind. That's why ESI consistently earns praise for **exceeding** customer expectations.

- **Financial stability and strength** — ESI has been self-funded and profitable since its inception. Effective management of the company's growth (in excess of 35% per year over the past five years) and sound fiscal policies are the guiding principles of the ESI management team.

Businesses are searching for suppliers whose products are designed to meet their specific needs. ESI continues to develop products that are significantly easier to use. ESI combines today's products with a clear vision of the future, and is prepared to grow with you — now and for many years to come.

ESI's slogan says it all: ***We make it easy to communicate!***

Maher Technology Group

Over the years I have had the pleasure of working with the owner of the Maher Technology Group – Joe Maher. The simple truth of the matter is this. As the two worlds of voice and data technology are converging upon us, it is getting harder and harder to distinguish between the two facets. What the Maher Technology Group does is specialize in the networking and integration of these two services.

MTG works with a variety of telephone systems, such as Northern, IVX, Toshiba, AT&T and a variety of others, as well as working with your current computer system, network, or building you a complete system from scratch. What MTG does better than anyone else, is ensure that your system works the way that it is suppose to, and stays working.

On more than one occasion, MTG has been known to work well past regular business hours. They have worked over weekends to ensure the proper installation of their products and to satisfy customers who could not afford any down time during weekdays. MTG has sound reliable service and a work ethic that is hard to match these days. Below is their company information.

MAHER TECHNOLOGY GROUP

Maher Technology Group is a full service data and voice integration company, offering telephony applications, data infrastructure, and consulting for today's growing businesses. The professionals at Maher Technology Group have an extensive background in innovative and cost effective methods for meeting clients' requirements in design, installation, and maintenance of sophisticated multi-protocol, multi-platform network systems.

Over the years the team at Maher Technology Group has grown a substantial customer base that ranges from large corporations to small businesses. We value the relationships that we have formed and appreciate that no matter how large or small a company is, the networking solution being provided is a critical element in the well-being and success of that organization.

Some of the services we offer:

*Voice over IP (VoIP)

*Computer Telephony Integration

*Enterprise networking

*LAN/WAN Configuration

*Infrastructure and Cabling

With customer satisfaction our number one priority, our entire staff is dedicated to excellence. Every team member is dedicated to delivering a quality product to our patrons. All aspects of a job are tracked and coordinated through a project management database to insure that every phase is carried through to completion with minimal delay.

Maher Technology Group strives to be a leader in data integration and networking. All of our technicians hold industry respected Microsoft Certified Systems Engineer certifications, so our customers can rest assured that experts will maintain their systems. "Maher Technology Group, comprehensive solutions for a connected world."

Maher Technology Group

17072 Silica Drive Suite 101

Victorville, CA 92392

Voice (760) 559-1100

Fax (760) 241-2011

Email solutions@mahertechgroup.com

Intelisys

I have been in the telecommunications business for over twenty (20) years. It seems that the one constant I have found in this business, is the sales persons ability to be screwed over by every single long distance telephone company going. I am not trying to be facetious; it just seems to be more often than not, all too common practice.

Here is how it traditionally works. You find a long distance vendor, either directly or indirectly, or perhaps a broker who works with several firms. You negotiate your best contract, and then merrily pursue business relationships with your customers in the areas of long distance, T1's, T3's and the like.

You will begin to sell, 1, 2, 10, up to 15 or so accounts, and then realize that your commissions do not seem to be coming in, either accurately, or worse yet – at all. In fact, some accounts have zero commissions, others are completely wrong – never wrong as in overpaid by the way, while other accounts seem to have no mathematical formula whatsoever to even attempt to figure out what has gone wrong. You complain, you wine, you yell, cajole, negotiate, escalate, do the work for them, get commitments, wait literally months for your just due rewards, all to discover that your hard earned commissions are never coming. So many things have gone wrong. They can't find your account, the carrier hasn't paid them, they can't find the customer, their deal has changed with the carrier, they haven't paid their bills, they're in chapter 11, they're either going or are out of business, and my personal favorite (which has happened twice to me) is that they have been bought out and the new company no longer recognizes your previous contract.

Since the above was all too familiar, I had literally decided to forgo the long distance side of the business, because it was just too exhausting, both emotionally and time wise. While firmly entrenched and quite happy in my initial decision, a customer of mine asked if I could do him a 'personal favor' and find him a good long distance carrier. Reluctantly I took on the assignment, and quite by chance I stumbled across a company by the name of Intelisys.

I scribbled my name on a contract and sold my customer, at a pretty good rate I might add, his new long distance contract for his massive $35 per month bill, which was running at the time around $60, so he was pretty happy. I had never given it a second thought, until about 60 or so days later when I received a check in the mail for $3 and some odd cents. I actually had to research through my files to discover where it had come from. As it turned out, Intelisys had paid me my just due commission, large or small, as it was owed, and on time! I was so taken back; I actually called my wife to tell her about this crazy company who sent me a check for just over $3. I decided to call this *Intelisys* company and find out a little bit more about them and their services.

As it turns out, the two owners Rick Dellar and Rick Sheldon were ex-sales people for other long distance companies. Oddly, they felt that in order to build their business, they had to pay their independent sales agents, the amount owed, for the correct account, and strangely – on the same day every month! You have no idea how long this took me to get use to. I actually receive e-mails a week in advance from Intelisys asking me to review my accounts ahead of time, as my commissions are being directly deposited into my account. I almost get giddy when I recount their accolades, but the simplicity of their industry practices combined with their unquestionable business ethics is just so unusual.

After further investigation I found out that the people at Intelisys were phenomenally organized, and that they actually cared about helping me sell my accounts. Intelisys is an extremely large broker for over twenty-eight (28) long distance carriers! They can, will and have assisted me in accounts ranging from $10 to $20,000 billed monthly. They cover local lines, local traffic, intrastate, interstate, intralata, DSL, T1's, T1's bonded, T3's, point-to-point T1's, fractional T1's, integrated T1's split between voice and data, frame relay, and the list just keeps on going. They are able to facilitate not just one quote, but multiple quotes for the same prospect. As well, they will allow you to work directly with their suppliers. YES, directly with their suppliers - unheard of – but true! Now, for this amazing service, you would think that their turn around time would be weeks. No, quite the opposite, in fact I receive a great many of my quotes the same day. Occasionally I'll get a quote the following day and on a few more complex matters, it can be a 2 or 3 day process, which is still quite exceptional.

I guess the bottom line is this. I can't find their weakness. They seem to do everything right, treat people professionally and try to motivate you to continually strive for excellence. What more could you possibly ask for? Quite frankly, if you are a company looking for great rates with the best possible deals available, you would be foolish not to find an Agent who works with Intelisys in your area and discover the difference. E-mail them at info@intelisyscorp.com to find an Agent in your area. If you are an Agent looking for a home, with a company that will actually pay you for your work, with a phenomenal selection of products and services www.intelisyscorp. com is where you want to begin and end your search. Finally, if you're a telco looking to sell your products or services, Intelisys will market your company accurately, professionally, and with a large number of very, very happy Agents! Below I have included a little blurb directly from Intelisys. I could not be happier with this company and I highly recommend them!

Intelisys

Independence. No limits.

Intelisys Facts

Intelisys, Inc. Headquarters

1318 Redwood Way, Suite 120

Petaluma, CA 94954

(707) 792-4900

Intelisys, Inc. Sales Office

2222 Martin Ave., Suite 270

Irvine, CA 92612

(949) 622-9561

Website: www.intelisyscorp.com   Email: info@intelisyscorp.com

Intelisys Overview

Founded in 1994 by telecommunications industry veterans Rick Dellar and Rick Sheldon, Intelisys, Inc. is one of the largest telecommunications master agency distributors in the United States, providing access to telecommunications products and services for over 500 Independent Sales Agent organizations and thousands of end-user customers worldwide.

Headquartered in Petaluma, California, Intelisys presently employs 30 full-time colleagues providing sales support, carrier training, and commissions support to its Agent Community. The company also has a full service sales office in Irvine, California that directly supports Agent's sales efforts with their end user customers.

Benefits of Agent Alliance with Intelisys

Intelisys' unique value proposition focuses, in part, on providing an exceptional customer experience to each independent Agent. Benefits of Agent alignment with Intelisys include:

 ♦ "Back office," SWAT team sales assistance and support, permitting Agents to focus attention on sales, marketing and servicing of new and existing end users (principally businesses), such as assistance with quotes, administrative and contract support;

 ♦ Broad, comprehensive and competitively priced product and service offerings from leading telecommunications carriers and suppliers;

 ♦ Intelisys' good will and contractual agreements with carriers and suppliers, and the ability to negotiate "Agent Protective" contracts with favorable pricing;

 ♦ Carrier and supplier extranets at the sub-agent level;

Benefits of Agent Alliance with Intelisys continued

 ♦ SLA's as short as 24 hours on pricing quotes;

- ◆ Intelisys' exceptionally well trained and experienced talent; and

- ◆ Direct deposit of monthly commissions.

Current Carriers and Suppliers

ACC Business	Primus
Access One	Qwest
Access Point	Qwest Local
ATI	Raindance
AT&T	SAVVIS
BellSouth	SBC-CA
EVault	SBC-Southwestern Bell
Global Crossing	Sprint
ICC	TelePacific
InfoHighway	TMC
Internap	TNCI
MegaPath	UCN
New Edge Networks	XO
PaeTec	Xpedite

Intelisys Team Resources

Intelisys provides Agent and Supplier Partners with support in many areas. The Intelisys Team includes a **Strategy Team** that works to facilitate growth in the business, provide Intelisys Agents with sales, and support experiences that are unmatched in the industry.

The Intelisys **Sales Overlay** and **SWAT Teams** provide Agents with support in the field, and are responsible for assisting Agents in achieving a higher close ratio on their sales opportunities. They assist with pre-qualification of Agent opportunities, and are available as a resource on calls with End-User-Customers, Suppliers, etc.

The Intelisys **Marketing Team** is responsible for making sure that Agents are aware of all relevant promotions, opportunities, and deals that they can find exclusively through Intelisys.

The Intelisys **Sales Support Team** provides Agents with quotes, pre and post sales assistance, and order provisioning. The Intelisys **Agent Development Team** provides training and support to Agents in understanding our portfolio and services, as well as in the resources available to them at both the Intelisys level and the Supplier level.

Commissions and Systems Support

Working in partnership with RPM Software, Intelisys has designed a web-based commission system to address the specific needs of our Agent Community, and to support our highest strategic objective of delivering an exceptional customer experience to all of our constituents. Looking ahead, RPM functionality will be extended to improve processes, communications, and sales opportunity management. Ultimately RPM will be a fully integrated system providing our Agents with quotes, order management, and complete supplier information.

Faxback

FaxBack is really an interesting company as they have managed to bring faxing into company's and organizations in a whole new way that allows for a great deal of flexibility and performance at the desktop level.

I worked with them very closely on a large deal involving a telemarketing firm that wanted to utilize fax broadcasting, fax on demand and to also integrate their system on to every desk-top while allowing for a back-up of every fax that came into the system visa vie the system administrator. What can I say, the system works flawlessly!

They have terrific sales and technical support, while being able to integrate into a variety of systems demonstrating their overall knowledge of the current environment and their innate sense of urgency with respect to overall customer satisfaction. Generally speaking, I, like others, do not enjoy calling in for tech support. You are usually stuck in some voicemail jail from hell with no way out. So it is no surprise that the company's that I work with are experts in satisfying the time conscious professional by offering state-of-the-art systems and technical staff to alleviate your tech support nightmares. FaxBack technicians are highly trained experts who can answer your questions and walk you through the system to get you up and running.

Simply put, between their high level of sales professionals (who actually know their product) and their tenured technical staff, FaxBack makes the grade as a company who can offer a great deal and backs it up with service and support.

They are a great company to work with and I highly recommend them. Below is their company information.

For nearly twenty years FaxBack has been a leading fax messaging company with solutions that radically simplify the way organizations communicate. We provide award-winning network fax servers, fax-on-demand, broadcasting and web-to-fax solutions that streamline information processes, get time-sensitive information into the hands of your audience faster than ever before while reducing the cost of doing business.

Our technology is enjoyed by thousands of global organizations including AT&T, Arco, Bank of America, Compaq, Kaiser

Permanente, Kodak, NEC, Sherwin-Williams and Wells Fargo. Countless other organizations in nearly every industry from real estate to manufacturing and travel to education and healthcare trust FaxBack and NET SatisFAXtion for their fax communications needs.

Backed by our software and years of industry expertise, companies can drastically reduce their costs, increase efficiency and position themselves to meet the business communication demands of today.

FaxBack's product line spans network fax server technology, broadcast software as well as Fax-on-Demand web-page-fax integration:

NET SatisFAXtion Small Business

Organizations can eliminate redundant and costly fax machines, and greatly improve worker productivity by bringing fax messaging capabilities directly to end-user's desktops. NET SatisFAXtion supports WinFax PRO as a desktop fax client FAXability, an easy-to-use browser-based fax solution for sending, receiving, and managing faxes. The Email Gateway allows users to send and receive faxes from popular email applications and is an optional add-on module. NET SatisFAXtion packages supports unlimited users and up to 4 ports of fax per server. Starting price is $595.00.

NET SatisFAXtion Broadcast Edition

Targeted at those company's involved in sending hundreds, or hundreds of thousands of documents quickly, efficiently and reliably. The NET SatisFAXtion Broadcast Edition includes such features as mail merge to personalize documents, automatic retry for busy or non-answering numbers, detailed reports to track activity, and "Do Not Fax" database list management. The product is scaled to support a maximum of 96 ports of fax per server (4 T1's per system). Starter systems include 4 ports of fax, NET SatisFAXtion fax server software and one FAXability Broadcast license.

NET SatisFAXtion Enterprise Suite

Aimed at mid-sized companies, larger organizations and enterprises, the NET SatisFAXtion Enterprise Suite is built on an open script engine and can support hundreds and even thousands of users concurrently. Ideal for companies that require multiple fax domains and fax servers, and mainframe ERP and BackOffice integration. Easily scalable and expandable, this Edition supports up to 96 ports of fax per server and unlimited users. Also included is; ActiveX and Connector API's, as well as Web Page Fax scripting technology for IT managers, webmasters and those involved in critical communications processes that want to add powerful fax functionality to their web sites or corporate intranets. The Enterprise Suite supports the widest range of end-user client options including FAXability, FAXability Broadcast, WinFax PRO as well as integration with popular email applications like Microsoft Exchange/Outlook, Lotus Notes and any SMTP Internet mail client.

NET SatisFAXtion Appliance

The NET SatisFAXtion Plug-and-Fax Appliance is a complete network server solution that includes the latest hardware technology from Dell along with our popular fax software. Network users can easily send, receive and manage faxes from FAXability, an easy-to-use browser-based application.

Fax-on-Demand

FaxBack Fax-On-Demand, blends automated voice response with fax server technology to provide a document delivery solution that runs 24 hours a day, 7 days a week. From a touch-tone telephone, your customers, prospects and associates follow a series of voice-prompted menus or "call paths" to access a catalog of documents, make selections and have them delivered to their fax machine within seconds.

For More Information

FaxBack, Inc. **Voice:** (503) 597-5350 or 1-800-329-2225

Warren Romanow

7409 SW Tech Center Drive, Ste. 100   **Fax:**   (503) 597-5399 or 1-800-329-6453

Tigard, OR 97223   **E-mail:** info@faxback. com

                                                    Web: www.faxback.com

NET SatisFAXtion was a recipient of the Windows .NE

Magazine Reader's Choice Award for network fax servers.

Identafone

There are a few things that I like about this product and this company. The IdentaFone software is utilized for screen synchronization, or 'screen pop' if you will.   The software is exceptionally flexible and not only works with a variety of business telecommunications products, but also with your home telephone as well.

Recently I was working on an installation integrating an IVX product with the IdentaFone software. What should have been reasonably simple was turning into a chore. The two software products, ESI's TAPI driver and the IdentaFone software were not seeing each other. What surprised me about IdentaFone was how accessible the owner was for consultation. I was able to call him well after normal business hours and have him run through a battery of system checks in order to identify the problem. In fact, he was even writing code on the fly in order to correct the integration conflict! Try to get that kind of service from any other provider – anywhere!

As it turns out, the problem was in ESI's TAPI driver. Once this was established, ESI corrected the problem by writing new software in less than a day! Integration established – problem solved, and the customer was very happy.

These are the kinds of products and company's that you want to surround yourself with, and IdentaFone certainly hits the mark! When you need screen synchronization – this is the company that you want to work with – bar none! I have included their contact information below.

IdentaFone Software

636 Pinerow Cres

Waterloo, Ontario N2T2K8

Canada

(519) 880-1214

URL: http://www.identafone.com

Email: cbw@identafone.com

Background

For nearly a decade IdentaFone software has been developing and deploying innovative CTI (Computer Telephony Integration) products. Our focus has been increasing telephone productivity on the desktop through screen pops and contact management integration. IdentaFone has products made for a wide range of hardware platforms including voice modems, specialized CID adapters and TAPI compliant IP phone systems.

IdentaFone is a small highly focused company based in Waterloo, Ontario, which is known for its technology innovations from locally started companies, such as RIM and Open Text. With a strong focus on customer and pre-sales support, we take the philosophy of problem solving as an inspiration for our software designs.

IdentaFone also provides custom CTI software development and is a supplier of software tools for Verizon Communications giving technicians the ability to conduct analysis from Gemini, Millennium and ProTel gateways.

IdentaFone's software products can be grouped by their hardware platforms with features aimed at two distinct types of users, namely the home user and the corporate desktop.

Products

IdentaPop Pro

Hardware Required: IP phone systems such as Cisco, ESI and NBX or any TAPI compliant phone system.

Features

- Unique display that overlays the CID on your screen and fades away.

- Can automatically pop a caller's contact in MS Outlook.

- Outlook integration works with public folders and Exchange Server.

- Automatically create a journal entry for incoming phone call.

IdentaFone MultiLine Software

Hardware Required: Multi line CID boxes from Zeus, Rochelle, YES Telecom or TCI.

Features

- A virtual telephone assistant that works with your multi-line CD hardware for monitoring from 2 to 100+ lines. The network version allows many workstations to have the same detailed status of callers and line usage.

- Full integration with MS Outlook Contacts.

- Reports and line usage statistics.

- Line usage display with timers.

- Downloads stored calls from hardware.

IdentaFone Pro

Hardware Required: A voice modem or TAPI compliant phone system.

Features:

- Speech announcements of Caller ID.

- Block telemarketers or any number you choose.

- Email, pager and mobile forwarding of CID.

- Speed Dialer.

- Custom messages played to callers.

Awarded the MSD2D.com 2004 People's Choice Award for Best Exchange

Connectivity Product at MS Tech Ed 2004 (San Diego)

Tapit

I have searched high and low for a flexible and compatible call accounting system. The truth of the matter is, that some work good with one product line, but not another. Another frustrating point is the level of service that you receive from the manufacturer. Personally, I don't believe that "Call the manufacturer of the telephone system" is the proper response to any question that I am posing to a call accounting company. Nor do I wish to sit on hold for the next 3 hours.

In comes TAPIT Call Accounting. This is a company that has not lost the personal touch of human interaction. Not only do they have a great product, they also call you to see how you are doing. Their technicians will call you back. Let me say that again. Their technicians will actually call you back! They really want to help you, and they really want their product to not only work, but to work the way it was designed to work!

TAPIT is uniquely structured so that it merely accepts what the telephone system spits out. In other words, directly from the SMDR (Station Management Detailed Recording) port each individual system will outsource information about the inbound caller, outbound calls, the number dialed, etc.

Some systems generate a wealth of information, and some systems generate very little. What is so nice about the TAPIT system is that whatever the system puts out, the TAPIT system will take that information and correlate it into a report. The more information your system generates the more reports you will receive.

I have tried their product on a host of different systems and have yet to be disappointed. The other fantastic benefit is that if you are not use to their software or running reports, you will still not be confused. The software is user friendly and bulletproof! You are going to love how easy it is to use, and how you can vary the operation of your company with the information that TAPIT can provide you with.

They also have a litany of reports, which will make even the most anal supervisor satisfied with the information and the quality of the statistical analysis incredibly happy. Below is a write up from the company.

Key Features Include:	Frequently Requested Reports Include:
• *Unlimited* extensions, departments, account codes and call records	• Longest calls by extension
• *Fraud Alert Detection* module	• Most expensive calls by extension
• *Web Report Viewer* for browser-based report access (network version)	• Most frequently dialed numbers
• *Automatic report scheduler* sent to screen, file, printer or EMAIL	• General Summary reports
• *Remote Manager* for Multiple Site Polling	• Account & Matter Code reports
• Extensive *pre-formatted reports* plus a custom report designer	• Departmental reports
	• Trunk utilization
• *Export* to Time and Billing packages, spreadsheet or text file	• Area code reports
	• Caller ID Deluxe reporting
• Transferred calls and station to station reporting	• Graphical reports
• Flexible *call costing* functionality	• DNIS reports
• Network Version allows report access for up to 255 simultaneous users	**Minimum System Requirements:** Pentium II or greater 128 MB RAM 250 MB free disk space Windows 98/NT/2000/XP Professional

Tapit EX™ stores call record information generated by your phone system enabling you to recall it in your choice of report formats. As a general business solution, **Tapit EX™** provides important information to help manage any organization. Try a demo today and see how **Tapit EX™** will control costs, increase productivity, better manage personnel, generate revenue, allocate calls to various cost centers, track advertising costs, identify fraudulent use…and more.

Tality™ is our software solution to the hospitality industry that offers a variety of capabilities to make phone activity a profitable part of your business. Generate detailed bills that include recurring charges, taxes surcharges etc. Run as a standalone package or export to ANY Property Management System.

Trisys, Inc. 215 Ridgedale Avenue, Florham Park, NJ 07932
973-360-2300 www.Trisys.com

Please contact Jason English ext.125, jasonE@Trisys.com for additional information.

Sureen Systems

When it comes to writing software and managing large systems, you'll find it a hard task to match the professionalism and high standards of Sureen Systems. They are bar none the most inventive group I have ever come across. Where there was no solution for your problems yesterday, they will craft one for you today.

I have been working with the owner of Sureen Systems, Sanjay, for quite some time now, and I have always found him to be my safe haven in a storm. He is inventive, and very methodical in his approach. He does not throw in all of the fluff that usually accompanies software development, but rather a straight line between two points that will get you where you want to be and your job completed.

Sureen Systems has a true understanding of IP connectivity and the integration that must be present in order for all of the integration to take place seamlessly. I would not hesitate one bit in recommending this company. Below is some additional information and references for work that they have done.

Sureen Systems

Please allow me to Introduce **Sureen Systems**, offering a comprehensive solution to your computer network needs. Sureen Systems has been in business in the UK for over 15 years, has "crossed the pond" so to speak, and is now proud to offer the same excellent level of IT / Computer services to businesses in the Los Angeles area.

Sureen Systems specializes in small to medium sized business like yours. We understand that staying at the top of your competitive market involves many things, and chief among them is your IT infrastructure, capabilities and how you employ them to your greatest advantage.

With our team of experienced technicians, who have been selected for their superior knowledge and skills, in addition to their troubleshooting / diagnostic capabilities. You are in good hands with the seasoned professional from Sureen Systems.

We will:

Troubleshoot existing problems within your company and solve them with the least amount of expenditure. Many times you don't need to scrap an existing network or equipment to gain a significant increase in productivity and efficiency. We are with you for the long haul and we will never recommend unnecessary equipment or software.

Blueprint your needs for today and the future. We will provide you with a roadmap, outlining your company's IT strategy that will increase efficiency and productivity immediately, while allowing for easy upgrades and expansion to accommodate future growth and trends in technology specific to your industry.

Implement a solution tailor made to your company's specific hardware, software, network and employee training needs. Sureen Systems can provide all hardware & software at a deep discount from retail prices. Using your custom "blueprint" we will put your system in place with little or no downtime and bring your employees up to speed with the training they need.

Support your network. We are always available to make sure your network is up and running by providing telephone support or an on-site visit from a technician. We know the latest technologies and how to best implement them to give you the edge over your competitors. A brief list of our services includes:

System repairs	**System Upgrades**	**Software Configuration**
Software Upgrade	**Network Installation**	**Website Design**
Website Hosting	**Training**	**T1 Installation**
Disaster Recovery	**Equipment Warranty**	**24/7 Support**
Security / Firewall	**Teleconferencing**	

Clients list.

Reuters

Provided upgrade path for a 300 user network from Novell to a dedicated fault tolerant NT environment, with full roaming ability to prevent user down time. Automated workstation rebuild and application deployment. Network infrastructure capacity planning and relocation, provide enterprise wide solutions. UAT of new thin client technology solution and implemented all over Europe. Centralized administration for all locations.

Universal Studios

Development server support, relocation of Data Center, Thin client application load testing for up to 300 client machines, IIS multi hosted support.

Re/Max

Consultation with the new owner to find out what their expectations and knowledge level was, of the technologies available. Provide documentation of required equipment, costs, etc. Worked with other out side resources to make sure that power, network and cabling was to customer's network specifications. Provide latest hardware configured with the required software to make this 'one of a kind' Re/Max office function to this specific requirements. All conference rooms were to be able to access the Internet and realtor specialist applications. All realtor agents needed to have a connection to the office T1 utilizing their own equipment. Security implantation had to be developed in order to stop any connectivity to sensitive data, with the latest tested and proven technology, providing agents with the latest solutions to help achieve their goals.

Legacy Escrow

Needed to set up and organize a centralized server and printing facilities to accommodate a large escrow with a huge volume of paper work. Needed to develop procedures for over night updates to franchise locations.

Aytel Systems Limited

Hosting solution provided to this pioneering web development solution provider, for its extensive customer base and their many requirements.

Conoco Limited

On site hardware support to ensure maximum up time for this multi thousand manpower location. Service Level Agreements for various hardware vendor equipment. Maintaining inventory of all on site hardware.

Royal Automobile Club (RAC)

Installation of Cisco router over the length and breath of the UK to centralize payroll processing for this 10,000 and month payroll run organization. Provide automated fail-over solution to another part of the country if the primary payroll location was not in operational.

American Coast Funding

Supply and installation of Windows 2000 server and desktop workstations. Providing this very dynamic fund specialist with an efficient, and low cost of ownership solution. Providing a centralized location for all data, to make contacts and information available uniformly to all employees.

Please take 5 minutes out of your busy schedule to discuss your needs with a Sureen Systems professional. Give us a call at **1-866-4-SUREEN** or visit our website at **www.sureensystems.com**.

Your Home Ip Based System

The Mediatrix® 2102 features voice prioritization over data and innovative IP technology to connect up to two residential phones or fax machines and a PC directly to a broadband modem - without the need for an external router. The Mediatrix 2102 VoIP access device is equipped with two FXS ports and two 10/100 BaseT Ethernet ports and is the ideal platform for service providers looking to deploy cost-effective residential IP telephony.

Protecting Your Investment

The Mediatrix 2102 can be deployed as a residential IP telephony access device, enabling service providers to cost-effectively deliver VoIP services to their subscriber base. The Mediatrix 2102 connects up to two analog phones and/or faxes, as well as a PC, to a service provider's network over a single broadband connection. Through a 10/100 BaseT Ethernet WAN interface, the Mediatrix 2102 connects analog terminals and a PC directly to a broadband modem without the need for an external router, and with only a single IP address provided by the service provider. With an embedded PPPoE client and its innovative IP technology, the Mediatrix 2102 and the PC connected to the second Ethernet port have the same public IP address, without private IP addresses or any address translation necessary.

Analog Series / 2102 - Residential VoIP Access Device

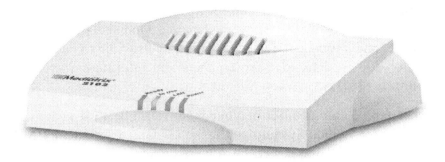

Key Features:

- IP connectivity for analog phones and faxes

- PSTN-quality voice over IP networks

- Deployable in SIP VoIP networks

- Automatic firmware and configuration updates

- TFTP or HTTP auto-provisioning

- User-friendly web interface

- Additional Ethernet port for PC or LAN connection

- PPPoE client

- Transparent IP address sharing

- Priority of voice over data

- Interoperable with equipment from leading industry vendors

- Fax over IP support, including T.38

- Multiple codec support

Applications

For service providers, the Mediatrix 2102 creates additional revenue-generating opportunities by immediately bringing residential users online to new high-value IP telephony services. Quite simply, the Mediatrix 2102 is an ideal, cost-effective solution for bringing VoIP to the home by using the existing broadband connection, and without having to invest in any additional devices.

This means that the Mediatrix 2102 is the only device that is needed in the home, in addition to the cable or DSL modem, to allow service providers to deploy residential IP telephony services.

With the Mediatrix 2102, service providers have the product characteristics that are needed to successfully deploy residential IP telephony applications. The Mediatrix 2102 provides a web interface, giving users a convenient access to the unit for initial set-up. The Mediatrix 2102 can auto-provision itself by fetching its configuration file from a TFTP or HTTP server making installation transparent to end-users. To further facilitate deployments, factory loaded configurations are possible. Automatic firmware and configuration file downloads ensure that the 2102 is always up-to-date.

Chapter Ten
Glossary of Terms

1) A through Z

Glossary

A

A-bit

One of the signaling bits used in Channel Associated Signaling. On a standard T1 two signaling bits are used (A and B). In other signaling protocols four signaling bits may be used (A, B, C, and D).

ACD

Automatic Call Distribution: An ACD is a specialized telephone switch, which provides queuing of inbound calls to pools of agents.

ADSL

Asymmetric Digital Subscriber Line: A digital line that uses the existing twisted pair copper telephone network to achieve speeds of up to 6 megabits per second up to 12000 feet, or 1.5 megabits per second up to 18000 feet.

ALAW

A form of voice encoding for digitized voice: This form of voice encoding is used in Europe. ULAW (or MU-LAW) Voice encoding is used in the United States.

AMI

Alternate Mark Inversion: A method on T1 digital lines of representing '1' bits by voltages of alternating polarity.

AMIS

Audio Messaging Interchange Specification: A protocol governing the exchange of voice messages and other information between voice mail and voice processing systems. This standard allows some degree of inter-operability between equipment from different vendors who support the protocol. Two separate AMIS protocols are defined - digital and analog.

Amplitude

The magnitude or strength of a waveform. The greater the amplitude, the greater the energy

Analogue (analog - US)

Pertaining to data that is transmitted in the form of a continuously varying electrical signal. The pitch and volume of the caller's voice is represented by the analogue signal.

ANI

Automatic Number Identification: A service, which passes the calling party's phone number to the receiving party during call setup. Often ANI is used to refer to any such service, which provides the calling party's phone number.

ARPANET (Advanced Research Projects Agency Network)

One of the earliest networks. It provided a vehicle for networking research centers and universities. ARPANET was the basis for the evolution of the Internet.

Area Code

The first three digits of a 10-digit telephone number. Designates a geographical area within which, station numbers are sub grouped. Three-digit code designating a geographical division within the North American Numbering Plan.

ASCII (pronounced as-key)

The American Standard Code for Information Interchange is a universally recognized format used mainly for 'text file' exchange. ASCII uses bits to represent different alphanumeric symbols (for example, ABC, xyz, !, @, +, ABC) and control codes (for example, tab, backspace, carriage return).

Asymmetrical

Providing differing bandwidth in different directions. 56 K modems are asymmetrical: they offer a maximum speed of 56K for downloading, but only 28.8K or 33.6K for uploading.

B

Bandwidth

The maximum units of data that can be transmitted per second through a channel. Measured in hertz in an analogue system and in bits per second in digital systems.

Baud

A measure of signal changes per second. Often used incorrectly in place of bps (bits per second).

B-Channel

Bearer channel: A circuit-switched digital channel that sends and receives data, voice, or video signals at speeds up to 64 Kbps.

Binary

Pertaining to a numbering system with a base of two (as compared to 10's in the decimal system), consisting of the values "0" and "1".

BPS

Bits per second.

BPV

BiPolar Violation: On T1 trunks, which use Alternate Mark Inversion (AMI) when two successive 'circuits's of the same voltage polarity are received.

BRI

Basic Rate Interface: A consumer grade ISDN line consisting of 2 64K bearer channels and one 16K delta (controller) channel.

Broadband

A high-capacity communications circuit/path. It usually implies a speed greater than 1.544 Mbps.

C

Caller ID

A protocol used to provide calling party information on a standard loop start phone line. This information is provided via a specified modem protocol between ringing signals.

Carrier - 1

A telecommunications provider, which owns its own switching equipment that it rents, leases or sells to the public for a set fee.

Carrier - 2

A radio wave, which is modulated by another signal for transmission over the airways (see also Modulation).

CAS

Channel Associated Signaling: Signaling is the exchange of call control information between the phone switch and the caller. With Channel Associated Signaling this information is included within each voice channel on a digital link. For example, DTMF tones may be passed in-band across the voice channel. The alternate method of Common Channel Signaling provides a single signaling path for a group of voice channels outside of the voice channel. Common Channel Signaling is used in ISDN trunks.

CCITT

International Consultative Committee on Telecommunications and Telegraphy: The CCITT acronym comes from the French Commite' Consultatif International de Telegraphique et Telephonique. An international standards body, known as the ITU-T since March 1, 1993.

Central Office (CO)

A telephone company switching center, in which is found a telephone switch that connects to customers' telephone lines.

Central Processing Unit (CPU)

The brain of a computer, which contains the circuitry that interprets information and executes instructions.

Centrex

A service offered by your local phone company, which provides PBX features (including the dialing of extensions for other local users). Unlike a PBX, the switching equipment is located at your Central Office, and is owned and maintained by your phone company.

Channel

A transmission path between two points. It is usually the smallest subdivision of a transmission system by means of which a single type of communications service is provided.

Channel Bank

The terminal equipment used to combine 12 or 24 voice channels together.

Cleardown

Many phone switches provide some tone to indicate that the remote party has dropped. In many situations the system can be configured to detect this Cleardown tone and end the call. Cleardown tones are not standardized and will differ between PBX/ACD vendors and between countries. This is also called a disconnect signal.

CLEC

Competitive Local Exchange Carrier: An alternative to the existing local phone company.

Common Channel Signaling (CCS)

A network of high-speed links connecting Digital Multiplex Systems (DMS). Information such as 'on-hook', 'off-hook', and telephone numbers are carried over common channels.

CO

Central Office: In reference to the phone company's central switching station for a given area, such as; a C.O. Line is simply a line (home or business) provided by the Central Office.

CO Central Office

The place where the Public Network switching equipment is located.

Coax

The abbreviation for coaxial cable.

Companding

A method of representing samples of digitized voice. Two standards of companding are in common use. The ULAW standard is used in North America and Japan. The ALAW standard is used in Europe.

Crosstalk

Undesired energy transferred from one voice circuit to another.

CSU

Channel Service Unit: A device, which is required by law in the United States when connecting a T1 device to the Public Telephony Network. The CSU protects the network and provides remote diagnostic and service capabilities.

D

D-Channel

A term used to describe a channel used to pass data during a telecommunications session. The "D" channel is used to carry control signals and customer call data in a packet switched mode. In the BRI (Basic Rate Interface), the "D" channel runs at 16,000 bits per second (16 Kbps), part of which will carry setup, teardown, ANI, and other characteristics of the call. In the PRI (Primary Rate Interface) the "D" channel provides the signaling information for each of the 23 voice channels, (referred to as "B" channels). The actual data which travels on the D channel is much like that of a common serial port.

Decoded

Converted (as in data) back to its original state, i.e. before having been encoded.

Demodulation

Converting analog signals back into digital signals. A modem is a Modulator / DEModulator.

D4 Framing

The standard protocol for providing framing across a T1 link used for standard switched telephony operation. The framing consists of a standard repeating bit pattern on the digital link, which allows each side of the T1 circuit to synchronize with each other.

Dial Table

A data structure used in the EWF to control the translation of a phone number into a dial string for placing outbound calls and transfers. For example, the dial table might specify that a 9 be added before external calls.

Dial Tone

A signal provided by telephony equipment to indicate that they are ready to receive dialed digits. Most PBX / Key systems provide dial tone after a line goes off-hook, or after a flash.

Diaphragm

A vibrating disk.

DID

Direct Inward Dialing: A line protocol which provides the dialed phone number as part of call setup. DID lines provide inbound service only.

Digital

Pertaining to data in the form of a sequence of ones and zeros (bits), which is stored and interpreted by a network.

DNIS

Dialed Number Identification Service: A telephony service which provides the dialed party's phone number as part of the inbound call setup. On E&M wink start lines this information is provided via in-band DTMF (Dual Tone Multi Frequency) digits and is a standard set of 16 tone combinations used in the USA and much of the world for dialing.

DS1

A high-speed line capable of delivering 1.54 Mbps (1,540K) in both directions, and divided into 24 data-bearing channels.

DS1C

A high-speed line capable of delivering 3.15 Mbps (3,150K) in both directions.

DS2

A high-speed line capable of delivering 6.31 Mbps (6,310K) in both directions.

DS3

A high-speed line capable of delivering 44.7 Mbps (44,700K) in both directions.

DSP

Digital signal processor. A specialized processor, usually for handling audio or video signals.

DSVD

Digital Simultaneous Voice and Data.

ECM (Error Correction Mode)

An option within the CCITT FAX protocol, which allows the retransmission of blocks of FAX data, if transmission errors have been detected. Both the sending and receiving FAX machines must agree to support ECM during initial T.30 handshaking or this feature will not be used.

Electro-magnetic Interference

Interference (noise) induced on a system by energy radiating from an electrical source such as a motor.

E&M WinkStart

A signaling protocol supported on T1 lines on telephony systems.

ESF (Extended Super Frame)

A T1 framing method used to uniquely identify the positions of the 24 timeslots for the 24 voice channels. ESF is a more sophisticated method than "standard framing", and provides additional signaling features.

E-1

Roughly the European equivalent of a T1 or a PRI, but with 30 data-bearing channels

E1 A

Digital telephony standard used throughout much of Europe.

Far End Disconnect

This term refers to methods for detecting that a remote party has hung up. This is also known as HangUp Supervision. There are several methods that may be used by a PBX/ACD to signal that the remote party has hung up, including Cleardown tone, or a wink.

Faxback or Fax-On-Demand

This is when the system receives an inbound call, and then later sends an outbound FAX on the same call. Often this FAX will be in response to user input.

FCC (Federal Communications Commission)

The US Government agency that regulates issues relating to telephony and use of the electromagnetic spectrum.

FIFO

First In, First Out: A type of data buffering that prevents data loss during high-speed communications.

Flash

A signal provided to a PBX/ACD to access special features such as transfers. On loop start lines a flash is performed by momentarily dropping loop current.

Four Wire

A connection that requires 4 conducting paths (wires) for a two-way communication. Two wires are used for send and two for receive.

Framing

A repeating bit pattern used on digital T1 and or E1 trunks to allow the two sides to synchronize with each other.

Frequency

Amount of times per second in which an electromagnetic wave completes a full cycle. One Hertz (Hz) equals one cycle per second.

Full Duplex

A communication path, which is able to carry information in both directions simultaneously.

G

Gamma, Link Gamma, Link Glare

This is when one inadvertently answers an inbound call when placing an outbound call. Glare is a problem on analog loop start phone lines, which are used for both inbound and outbound calling. Many phone lines using line protocols other than loop start are not subject to glare, such as E&M, Wink Start and ISDN, etc.

Gateway

A combination of hardware and software that bridges two different communications networks, permitting users on each network to exchange information.

GHz (Gigahertz)

1,000 MHz-one billion hertz.

Ground Start

A signaling method used on analog lines.

Group 3

One of the four sets of international standards existing for FAX transmission. Group 3 is the standard used by nearly all-standard FAX machines around the world. Group 1 and 2 FAX machines are obsolete. Group 4 FAX machines are expensive, and not yet widely available. Most new systems support Group 3 FAX operations.

H

Half Duplex

A communication circuit, which can carry information in both directions, but only in one direction at a time.

Hang up Supervision

Signaling provided by the phone switch to the telephony system to indicate that the remote party in a phone call has hung up. Such hang up supervision allows the telephony system to immediately free the phone line for another call. Different mechanisms exist to signal hang up supervision, depending on the line protocol used. On analog, loop start lines often a "wink" or momentary loss of loop current is sent to signal hang up supervision. Other systems may provide special tones known as Cleardown Tones to signal remote hang up.

Hertz (Hz)

Named after the German physicist Heinrich R. Hertz, a unit of frequency equal to one cycle per second.

Hierarchy

Data arranged in an organized series consisting of graded levels.

Hook Flash

A momentary depression of the switch hook to alert a PBX or switch, but not so long as to signal a disconnect.

Host

A centralized computer, which supplies data to PCs on a network, or a centralized telephone switch, which provides switching services to several smaller remotes.

HSP Host signal processor

In modems, a modem that depends on the host CPU (the Pentium, PowerPC, etc., in the main computer) for part or most of the data processing.

Hubs

A device used to centralize where all nodes are wired in a LAN.

Hunt Group

A pool of phone lines organized together for a single application. The routing of incoming calls to members of the hunt group may be controlled in various ways. On simple systems calls may be routed via a single pilot number to the next available number. On complex ACD systems, the routing of calls may be controlled by more sophisticated algorithms.

Hybrid

A device used in communication networks that converts a four-wire voice circuit into a two-wire circuit.

I

Inband Signaling

The exchange of signaling information for a call, inside the voice path of a call. For example, the digits of the desired phone number for an outbound call might be passed as DTMF digits within the voice path. Likewise, ANI/DNIS digits might be passed as DTMF digits in the voice path of an inbound call.

Interface

A common boundary between two pieces of equipment where they join together, enabling them to exchange information.

ISDN

Integrated Services Digital Network: A digital phone service capable of speeds from 57.6K to 128K, which provides two data channels, each with its own phone number making simultaneous voice and data possible.

ISP

Internet Service Provider: A company that provides access to the Internet through modems, ISDN, T1's, etc.

ITU

International Telecommunications Union: An international standards body. Known as the

IVR

(Interactive Voice Response) Equipment, which performs applications involving callers on the phone utilizing both touch tone and voice recognition to have the system being used to interact with the caller in a way that would provide specific information as requested by the caller.

K

Kbps

kilobits per second: Kbps is kilo*bytes* per second.

LED

(Light Emitting Diode) These are small devices used on PC hardware, which can be illuminated. Many of the hardware cards supported by the EWF may provide LED's to signal information, which can help diagnose problems. The field notes for all telephony/voice cards will include any LED's, which are available.

Line Protocol

The protocol used on a phone line between the telephony system and the switch to provide call setup, breakdown and signaling. Examples include loop start, ground start, E&M WinkStart, and DID.

Line Type

A data structure used in the EWF to identify the capabilities of one or more phone lines.

Local loop

The copper wires running between the telephone subscriber's home or business and the phone company switch.

Logical

The electronics involved with adding/subtracting 1's and 0's is called binary logic. A "1" or a "0" generated by the electronics is called a logical 1 or 0.

Loop back

Also known as loop back transfer. Loop back is when two phone lines are connected internally within the telephony system. This mechanism may be used to allow a caller to be transferred on systems, which do not have a transfer capability. It has the disadvantage of requiring that the telephony system hold both phone lines while the two parties communicate.

Loop start

A common line protocol used on analog phone lines.

M

MAC - (Medium Access Control)

Internationally unique hardware identification address that is assigned to the NIC (Network Interface Card), which interfaces the node to the LAN.

Mbps

Megabits per second: MBps would be mega*bytes* per second.

Modem

A MOdulator/DEModulator. A device that can encode digital signals from a computer into analog signals that can be transmitted over analog lines, and vice versa.

Modulate

To vary the amplitude, frequency or phase of a carrier wave in order to transmit information.

Modulation

Converting digital signals into analog signals. A modem is a Modulator / DEModulator.

Multiplex

To transmit more than one message, at one time, on a single communications channel.

N

Network

A framework of several telephone switches which together, permit seamless transmission of telephone calls.

Network Processor

A centrally located computer, which monitors national voice traffic.

NSFnet (National Science Foundation Network)

A high-speed network that forms part of the Internet backbone.

O

Octet

A byte composed of eight binary characters (bits).

Off hook

Literally this refers to the state of a standard loop start phone line when the phone receiver is lifted up (off of the hook position). In this state the phone can dial and be connected to other parties. This term is typically used in a more general fashion to indicate when any phone line is in a state where it can signal to and from the phone system and be connected to other parties.

OPS

(Off Premise Station) A family of protocols used on T1 digital lines, which emulate standard loop start signaling. OPS lines are often the only way to get transfer capabilities on T1 lines. There are several flavors of OPS protocols.

Optical Carrier (OC)

The speed rate of an optical transmission, according to the SONET standard.

Orphan Call

An inbound call with no application to receive it. For example, if you do not have any inbound applications installed, then any inbound calls will be orphan calls. These calls will just ring until the caller gives up or the PBX/ACD forwards them.

On hook

The state of a phone line when it is idle with no call in progress. For a standard telephone this is when the receiver is hung up so that no connection exists.

OC-3

A fiber optic line capable of 155 megabits per second (155,000K).

OC-48

A fiber optic line capable of 2400 megabits per second (2,400,000K).

P

Packets

An accumulation of 'bits', made up of data and control information, which is grouped together and treated by the network as a single unit. The Packet is sent by one node to another across the network. The term packet and frame are often interchanged.

Peripheral

With respect to the telephone switch, it is any equipment connected to, but not in the same building as the local switch.

PBX

(Private Branch eXchange) A private phone switch, which is used to connect a pool of local phones to the public Network. PBX's often provide sophisticated features for call control, telephony and data operation.

PCM

Pulse Code Modulation: A method of encoding an audio signal in digital format.

PCMCIA

Personal Computer Memory Card International Association: A standard for miniaturized laptop expansion cards for modems, storage, and other devices, often called PC cards.

Phase

A variation in a signal, measured in degrees, from one reference point to another.

Pilot Number

A single phone number, which is used to route calls to a number of destinations within a hunt group.

POP

Point of Presence: A local dialing point for an Internet Service Provider.

Port

This is a term, which may have a variety of meanings in different contexts. To the Voice Subsystem this refers to a software, and, or hardware entity which provides services to a phone lines. These services include voice, text to speech, Voice Recognition, FAX, pulse detection, and DTMF detection and generation.

POTS

Plain Old Telephone Service: Regular analog phone service, as opposed to ISDN, ADSL, SDSL and other digital phone services.

Predictive Dialing

This is a service where telephony equipment (such as PBX, ACD, or IVR) dials outbound calls, and only connects them to a live agent when the call is successfully answered. This increases ACD agent productivity by allowing them to avoid handling calls, which are busy or Ring No Answer.

PRI

Primary Rate Interface: An industrial grade ISDN line. In the United States and Japan, a PRI consists of 23 64K bearer channels and a 64K delta (controller) channel. In Europe, a PRI consists of 30 bearer channels and a delta channel.

Prompts

The prompts (or voice prompts) are the sets of files needed to control the speaking of objects such as dates, numbers, currency amounts, and phone numbers. Each Language (such as American English, Italian, American Spanish, etc.) must provide its own prompt set.

PSTN

Public Switched Telephone Network:

Pulse Code Modulation

A four step process that converts an analogue signal to a digital signal by sampling the signal, quantizing it, encoding it, and multiplexing it with many other signals. The signal is converted back to its original analog state at the receiving end.

Pulse Signaling

A method of providing digits to the phone switch. This method is also sometimes known as rotary dialing, as older phones would generate these pulses by rotating the phone dial across a set of electrical contacts. Pulse signaling has been widely replaced by DTMF signaling in the USA, but is still in wide use in many countries of the world.

R

Regenerated

The restoring of a bit, which has been degraded by transmission impairments, back to its original form.

RBOC

Regional Bell Operating Company.

Ring No Answer

When a call fails after several rings without being answered.

RJ11

A standard connector for providing a single analog phone line.

Routing

The assignment of the preferred path for information to travel in order to reach its destination on a transmission network.

Rotary or Rotary Phone

The providing of phone number digits to the phone switch using pulses generated by the rotating of the phone dial across electrical contacts. This type of signaling has been mostly replaced by DTMF signaling in this country, but is still in wide use around the world.

RJ14

A standard connector, which provides two analog phone lines on a single connector. Physically the RJ14 is the same as the RJ11 except that two of the three available pairs of wires are used.

RJ48C

A standard telephony connector often used for T1 connections. The DTI48 T1 board from Natural Micro Systems provides two DTI48 connections. See the field note on DTI48 Installation & Troubleshooting for details, including a pin diagram of the connection.

S

Service Bureau

An enterprise, which maintains large sets of equipment which they use to sell specialized services to others. For example, a FAX Service Bureau might provide fax back service to others who do not want to support their own FAX hardware.

Slip

A loss and regaining of framing on a T1 trunk. Possible causes include:

Data errors on link cabling problems (loose or pulled cable) problems with clocking between sender or receiver cause loss of synchronization.

Subscriber/Termination

A common eight wire bus to which all devices connect.

Symmetrical

Providing equal speeds in both directions. Compare with asymmetrical.

Synchronized

Referring to a communications transmission system where input and output signals are timed at intervals that keep them operating in step with one another.

Synchronous

Occurring at the same time with regard to transmission systems, all locations are running off the same clock source.

Switch Equipment

The equipment, which connects (or switches) phone calls from a source to a destination. PBX's (Private Branch Exchanges) are phone switches, which are owned by the user and are typically onsite. ACD (Automatic Call Distribution) are specialized switches, which route inbound calls to pools of agents.

Switching System (or switch)

Communications equipment upon which each user has a unique address represented by his or her phone number. In response to a telephone number, the switching system or switch, selects the transmission path or circuit used to connect one user to another.

T

Talk off

When someone's speech is mistakenly identified as a touch-tone, causing an unwanted action to be selected. The EWF uses very reliable DSP-based algorithms for DTMF detection and has extremely low occurrences of talk off.

TAPI

(Telephony Access Protocol Interface)

Also known as Microsoft/Intel Telephony API. TAPI refers to the Windows Telephony API. TAPI is changing regularly to meet integration needs. Its current release is TAPI 2.0. TAPI simplifies the process of writing a telephone application that works with a wide variety of modems and other devices supported by TAPI drivers. In contrast to TSAPI (Telecommunications Systems Application Protocol Interface), TAPI places the emphasis on the Windows operating environment, while TSAPI operates at the Network Operating System level.

TCP

(Transmission Control Protocol) Protocol used on Internet and IP open architecture to transfer data across a network or wide area network.

Terminal Adapter

Electronic interface that makes non-ISDN devices look like ISDN.

TIFF

(Tagged Information File Format) A standard for representing compressed FAX images. The standard defines different types of TIFF files. In addition, vendor specific differences often exist between files used by different systems.

Time-division multiplexing (TDM)

A method of combining and transmitting several digital signals over a single line.

Tip & Ring

Standard telephone connections require 2 wires. For historical reasons relating to the manual plugs used by operations for switching, these 2 wires are known as Tip and Ring.

Token Ring

A type of LAN with nodes connected in a ring. Each node constantly passes a "token" or control message on to the next one. Only the node with the token can send a message.

Topology

Describes the physical layout and interaction of different facilities and services.

Transmission

The process of sending data from one place to be received at another.

Trunk

A high-capacity communications connection between two switching systems that provides outgoing, incoming or both, service to telephone subscribers.

TSAPI

(Telephony Server Access Protocol Interface)

An API developed by Novell and AT&T for control of telephony features. AT&T and Novell created the TSAPI standard to help pass controls between the PBX and a Novell server. AT&T describes TSAPI as "standards-based API for call control, call/device monitoring and query, call routing, device/system maintenance capabilities, and basic directory services." In contrast to TAPI, the control is at the NOS layer, not at the Windows operating system layer.

TTS

(Text To Speech) The ability to translate text into spoken voice. Twisted Pair Standard, inexpensive cabling commonly used for telephony applications. Typically it consists of two copper wires twisted around each other to reduce outside interference.

T-1

In North America, a digital carrier for a DS1-formatted signal.

T-3

In North America, a digital carrier for a DS3-formatted signal.

U

UART

Universal Asynchronous Receiver-Transmitter: A controller chip that processes data coming in and going out of the modem. The 16550 is a modern example.

ULAW

A method of companding (or compressing) digitized voice. This method is used in the United States, Canada, and Japan. The other method, ALAW, is used in Europe and much of the rest of the world.

V

Voice Mail

A specialized voice processing application, which is dedicated to accepting, controlling, and routing voice messages.

Voice Subsystem or Voice/FAX Subsystem

The portion of the system, which controls the real-time operation of phone lines and related technologies, including FAX, and Voice Recognition.

VPC

(Voice Process Corporation) A manufacturer of Voice Recognition hardware and software.

Voice Processing

This is the generic term for applications running on VRU and IVR systems. VRU (Voice Response Unit) is the generic term for equipment, which can perform automated call processing. Early VRU's did little more than answer calls, play prompts, and detect any DTMF tones entered by the caller. Advanced VRU's greatly enhance VRU usage by adding greater connectivity and more flexible application design.

V.32terbo

AT&T's proprietary protocol for 19200 bps asynchronous communications.

V.32bis

The ITU standard for 14400 bps modulation.

V.34

The ITU standard for 28800 bps and 33600 bps modulation.

V.90

The ITU standard for 56K modulation.

V.FC

Rockwell's proprietary protocol for 28800 bps asynchronous communications.

V.flex

Lucent's proprietary protocol for 56000 bps modulation. Merged with Rockwell's K56Plus to create K56flex.

W

WAN

Data communications network that serves users across a broad geographic area.

Wink

In general a wink is a type of signal sent across a phone line between two pieces of equipment. On an analog, loop start phone line, a wink consists of a momentary dropping of loop current. PBX's often use such winks to provide hang-up supervision signaling to the telephony system that the remote caller has dropped the call and that the line can be freed.

Z

ZCS

(Zero Code Suppression) A method used to ensure sufficient one's density on a T1 connection. On T1 circuits one must guarantee a certain density of voltage transitions to maintain clock synchronization between the sender and receiver. ZCS if used to insert a code containing a one when ever a string of too many zero's is detected. Obviously both the sender and receiver must agree that ZCS is being used to make sure that the original bit pattern is restored.

Chapter Eleven
References

References

White Paper – "An Introduction to IP Telephony" - Mockingbird Networks

IP Telephony OSS for Real-World Success - Eftia OSS Solutions Inc.

Technology Note – Dedicated Access Solutions – Dave Barone, Juniper Networks Inc.

IP Telephony overview & applications - Todd Compagna, Cisco

Voice Over IP Fundamentals - Jonathan Davidson, James Peters

Voice Over IP – The International Engineering Consortium

VoIP - Building a Sound Foundation for Voice over IP - Jason Lackey, Riverstone Networks

White Paper - Addressing Convergence and IP Telephony in Enterprise Communications, Iwatsu America, Inc

Technology Brief - Voice Over IP (VoIP) – Silicon Press

TCP/IP Version 3 Release 2 for MVS: Customization and Administration (SC31-7188)

VM/ESA V2R4 TCP/IP Planning and Customization (SC24-5847)

Packet Policing on the Passport 8600; Nortel Networks. 2000

Scalable High Speed IP Routing Lookups; White Paper/Academic paper

Broadband DVB-based Internets and Intranets via satellite; Siegfried Dickhoven & Olaf Menzel; German National Research Center for Information Technology

Evolution of 2-way IP Connectivity; Penny Glover Astra 2000

"Always on: Broadband Living Enabled." Broadband Innovation Group, MediaOne Labs, October. Anderson, Ken, and Anne Page McClard. 1998.

Chapter Twelve
Bibliography

Bibliography

"The Broadband Revolution: How Superfast Internet Access Changes Media Habits in American Households," October 2. Arbitron/ Coleman (presented by Pierre Bouvard and Warren Kurtzman) 2000

Broadband Access: Opportunities in Fixed Wireless. Surrey, U.K.: ARC Group. ARC Group. 2000.

Ramachandran Ramjee, Jim Kurose, Don Towsley, and Henning Schulzrinne. Adaptive playout mechanisms for packetized audio applications in wide-area networks. In Proceedings of the IEEE Infocom, pages 680-688, 1994

VoIP Products, services and issues, Cis.ohio-state.edu

Voice Over IP: Protocols and Standards, Cis.ohio-state.edu

Voice over IP [Audio/Video recording], Cis.ohio-state.edu

Voice over IP: Issues and Challenges, Cis.ohio-state.edu

Voice over IP [Audio/Video recording], Cis.ohio-state.edu

Voice and Telephony Over ATM, Cis.ohio-state.edu

H.323 and Associated Protocols, Cis.ohio-state.edu

An Introduction to IP Telephony, Mockingbird Networks

An Overview of Circuit, Packet, and SS7/C7 Networks, Mockingbird Networks

Applications for Internet Telephony, Dialogic

Ascend Voice Over IP - Frame Relay Resource Center, Alliancedatacom

IP Telephone Design and Implementation Issues, Telogy

IP Telephony Tutorial, Performance Technologies, Inc

IP telephony, Teledotcom

ITworld.com - VOIP - Internet telephony, voice over IP,

VoIP Howto, Linuxdoc.org

VoIP Psychology, Networkmagazine.com

VoIP in the Enterprise, Networkcomputing.com

Uyless D. Black, " Voice Over IP (2nd Edition)," Prentice Hall, January 2002, 400 pages.

David J. Wright, " Voice Over Packet Networks," Wiley, March 2001, 270 pages

Peter Loshin, "Big Book of IP Telephony RFCs," Morgan Kaufmann Publishers, January 2001, 784 pages.

Oliver C. Ibe, " Converged Network Architectures: Delivering Voice and Data Over IP, ATM, and Frame Relay," Wiley, November 2001, 416 pages.

White Paper – "An Introduction to IP Telephony" - Mockingbird Networks

IP Telephony OSS for Real-World Success - Eftia OSS Solutions Inc.

Technology Note – Dedicated Access Solutions – Dave Barone, Juniper Networks Inc.

IP Telephony overview & applications - Todd Compagna, Cisco

Voice Over IP Fundamentals - Jonathan Davidson, James Peters

Voice Over IP – The International Engineering Consortium

VoIP - Building a Sound Foundation for Voice over IP - Jason Lackey, Riverstone Networks

White Paper - Addressing Convergence and IP Telephony in Enterprise Communications, Iwatsu America, Inc

Technology Brief - Voice Over IP (VoIP) – Silicon Press

About The Author

Warren Romanow has been in the interconnect industry since 1984, selling, distributing and networking telecommunications equipment. He constructs systems inclusive of; telephones (IP), voice mail systems, unified messaging, screen synchronization and database look-up, facsimile networks, LAN/WAN, data and desk top environments, and pricing for local and long distance services.

He has held the positions of, Vice President of Telrad Canada, President of the Comtel Data Group and C.E.O. of Comtel International. Warren has worked with companies and installed telecommunications equipment and services in Canada, the United States, Egypt and Saudi Arabia.

Warren Romanow is now residing in Alta Loma, California where he owns and operates DataNet Telesys Group. DataNet specializes in seamless IP telecommunications, system networks, software integration, screen synchronization, innovative line restructuring, facsimile networks and cost justifications. Warren is a member of the International High IQ Society (membership number 3378 / I.Q. of 138) and is extremely proficient in customer cost analysis and often pioneering new telecommunications strategies.

There are four objectives in writing this book. 1. If you want to know from Alexander Graham Bell up until today, Warren explains it in an easy to understand language. 2. Actual real life examples of customers who should use IP telecommunications. 3. How to actually put the entire network together making it run smoothly and seamlessly. 4. Who to use to construct and achieve this entire network extraordinaire and why.

Warren explains, "I thought it best to put together a guide for the average business to follow. Roughly 86% of all business' fall into the small to medium size range, and although their requirements vary, their most basic needs remain the same. I let individuals, businesses, and vendors know whom they should deal with for the bulk of what they need. That is, if you want to sleep at night."